ADAM'S TONGUE

ADAM'S TONGUE

HOW HUMANS MADE LANGUAGE, HOW LANGUAGE MADE HUMANS

DEREK BICKERTON

HILL AND WANG

A DIVISION OF FARRAR, STRAUS AND GIROUX

NEW YORK

Hill and Wang
A division of Farrar, Straus and Giroux
18 West 18th Street, New York 10011

The Library of Congress has cataloged the hardcover edition as follows:
Bickerton, Derek.
    Adam's tongue : how humans made language, how language made humans /
Derek Bickerton.— 1st ed.
        p.   cm.
    Includes bibliographical references and index.
    ISBN: 978-0-8090-2281-6 (alk. paper)
    1. Language and languages.   2. Human evolution.
3. Psycholinguistics.   I. Title.

P106.B4667   2009
401—dc22

                                                              2008045906

Paperback ISBN: 978-0-8090-1647-1

Designed by Jonathan D. Lippincott

www.fsgbooks.com

1   3   5   7   9   10   8   6   4   2

CONTENTS

ADAM'S TONGUE

INTRODUCTION

You *can* try this at home.

No specially constructed course, safety equipment, or medical assistance of any kind is required.

It's what's called a thought experiment. Thought experiments are vital to science. Without thought experiments, we'd never have had the theory of relativity. If Einstein hadn't imagined what riding on a beam of light would be like, or what would happen if two gangsters shot at each other, one in a moving elevator, one outside, we'd still be stuck with the Newtonian universe.

This particular thought experiment is very simple. You just have to imagine for a moment that you don't have language and nobody else has, either.

Not speech, mind you. Language.

For some people these are synonymous. My heart sinks every time I open a new book on human evolution, turn to the index, and find the entry "language: *see* speech." "You don't see speech, you idiot," I feel like yelling. "You *hear* speech." You can have speech without it meaning a thing; lots of parrots do. Speech is simply one vehicle for language. Another is manual sign. (I'm talking about the structured sign languages of the deaf, like American Sign Language, not the ad hoc gestures hearing people use.) Language is what determines the meanings of words and signs and what combines them into meaningful wholes, wholes that add up to conversations, speeches, essays, epic poems. Language goes beyond that, even; it's what makes your thoughts truly meaningful, what builds your ideas into structured wholes. (If you doubt this, or feel it's a stretch, just read on to the end of the book.) Even if you think you think in images, language is what puts those images together to make meaningful wholes, rather than just disordered, tangled messes.

Think of how, without language, you'd do any of the things you do, without thinking, every day of your life. Writing letters (e-mail or snail mail). Answering the phone. Talking to those you love. Following the instructions for assembling the new gadget you just bought. Reading road signs (okay, some are graphic symbols, but the meanings of such symbols are not transparent—you have to learn, through language, that a picture of something with a diagonal line through it means you're not supposed to do that thing). Playing any game (you learned the rules, spoken or written, through language). Shopping (you couldn't read the labels on the cans; indeed, there wouldn't be any labels to read, if there was even a store to shop in). Rehearsing the excuses you'll make to your boss for coming in late. The list goes on and on. When you get to the end of it, here's what you'll find: everything you do that makes you human, each one of the countless things you can do that other species can't, depends crucially on language.

Language is what makes us human.

Maybe it's the only thing that makes us human.

It's also the greatest problem in science.

You don't agree with that? Well then, what would you say were the greatest problems in science? How life began? How the universe began? Whether there's intelligent life anywhere else in the universe? None of these are questions we could even ask if we didn't have language. How we got language is a question that logically precedes all other scientific questions, because without language there wouldn't be any scientific questions. How can we know whether our answers to those questions have any validity, if we don't even know how we came to be able to ask them?

Since the dawn of time humans have wondered what it means to be human. Every answer you can think of has been proposed, and some you couldn't have thought of. Plato defined humans as featherless bipeds and Diogenes refuted him with a plucked chicken. In 1758 Carl Linnaeus, the Swedish botanist who first classified species, termed us *Homo sapiens*—wise man—and later on, when the multiple-branching tree of human evolution was revealed, and we had to disentangle ourselves from Neanderthals and "early" *Homo sapiens* (presumed ancestor of both us and them), we became *Homo sapiens sapiens*: wisest of the wise. (Look around you and tell me if you think that's accurate.) Look up "human being" in the online *Encyclopaedia Britannica* and you'll find "a culture-bearing primate that is anatomically similar and related to the other great apes but is distinguished by a more highly developed brain and a resultant capacity for articulate speech and abstract reasoning." "Resultant," indeed! This is one of those remarks that seem to make sense, like "The sun rises in the east," until you ask yourself, is that what really happened?

Darwin knew a century and a half ago that the *Encyclopaedia* had it backward—that it wasn't a "highly developed brain" that gave us language (not speech!) and abstract thought, but language that gave us abstract thought and a highly developed brain. "If it be maintained that certain powers, such as self-consciousness, abstraction etc., are peculiar to man, it may well be that these are the incidental results of other highly advanced intellectual faculties, and these again are mainly the result of the continued use of a highly developed language."

Nobody followed up on this. It was bad enough having an ape as one's great-granddaddy—worse still if all that really divided us was that we could talk and he couldn't. It was much more flattering to our self-esteem to suppose that our marvelous brains and minds just . . . grew somehow, got smarter all by themselves, and then started pouring out a cornucopia of thought and invention, science and literature, all the things that proved us the wisest of the wise. So we heard endlessly that what distinguished us as humans was our consciousness, our self-consciousness, our foresight, our hindsight, our imagination, our ability to reason and to plan, and on and on. Not one word about how any of these miraculous abilities evolved. That might have forced us to really look at language and how language began and what it did for us. But the belief that language was merely one of many outputs of our wonderful brains, though not quite universal, was widespread enough to make language origins look like an isolated problem, one you could split off from the rest of evolution, even the rest of human evolution, and crack at leisure, when there wasn't anything more pressing to be done.

One thing writers on language origins all-too-often ignore, but that I want to emphasize throughout this book, is that language evolution is part of human evolution, and makes sense only if considered as a part of human evolution.

Another thing that discouraged people from coming to grips with language evolution was that it was such a hard problem. Insoluble, some said. In 1967 the psychologist Eric Lenneberg published a book, for the most part excellent, called *Biological Foundations of Language*. Now you'd think in a book with that title there would be, somewhere, some hint or at least a guess as to how those foundations got founded—how the mills of biological evolution managed to grind out such a unique product. But there isn't: Lenneberg concluded (always a rash move in science) that here was a question that could never be answered. Even two students of language evolution, writing very recently, described the origin of language as "the hardest problem in science." Language leaves no fossils. You can't do experiments (at least not ethical ones). Language is a population of one, a truly unique trait. And that's something all scientists dread, because it means you can't use comparative methods, and comparing things that are similar but differ slightly from one another forms one of the most fruitful procedures known to science.

It's hardly surprising, then, that attempts to explain how language evolved—and there's been a growing number of these over the past few years—should have gone off in dozens of different directions. Nor is it surprising that these explanations have shied away from the very heart of the problem. You can read endless accounts of what skills and capacities our ancestors had to have before they could get language, or what selective pressures might have favored the emergence of language; you can read accounts, not quite endless and usually sketchier, of how language developed once it had begun. But you will read little, and that little extremely vague, about what I once called "the magic moment"—the moment when our ancestors first broke away from the kind of communication system that had served all other species well for at least half a billion years.

How language evolved isn't just a hard problem in itself.

It's been made much harder to solve by two factors, both of which are actually quite irrelevant to it, but which we'll have to deal with if we are to start with a clear idea of what the problem really is (and also of what it *isn't*). One factor concerns the way in which evolution in general, and therefore human evolution in particular, has been presented by the neo-Darwinian consensus of the last century. I'll get to that in a moment. First I want to deal with an issue that will seem to many, perhaps most, as even more pressing and urgent: the status of the human species itself.

What's that got to do with language evolution?

You're right—nothing. And yet the evolution of language has been dragged willy-nilly into the culture wars, the epic and still unresolved struggle between those who want things to stay the way they are and those for whom they can't change too quickly.

Before the last century, there weren't many who dissented from the established view of "man's place in the universe." The human species, always identified with one half of it, was somewhere between ape and angel, equipped with an immortal soul (unlike the beasts), destined for eternal life (unlike the beasts), and in general enjoying an exalted status as a one-of-a-kind, specially created darling of the Almighty. Needless to say, the intellectual (as well as moral) powers of these anointed beings outshone the capacities of mere animals as the sun outshines the moon.

As Darwin's ideas spread, this notion of human status grew less and less sustainable. There gradually emerged an alternative view of humanity, humans as a species of ape, ground out like all species by the mills of natural selection, with nothing that made it more valuable than any other species, and nothing of any real importance that made it significantly different from any other species.

At first, this view served as a highly salutary corrective to the supremacist take on humans. But soon the two views were in full-on combat mode. And in war, if truth is the first casualty, objectivity goes out in the very next body bag.

There was an agenda (Get rid of superstitious nonsense!). There was a dogma (Evolution was always and everywhere a very slow and gradual process). On the rational scientific side (that's the godless materialist side, if you're on the other side), agenda and dogma combined to give a single program. It became mandatory to deny every difference between humans and other species that could in any way be interpreted as showing the superiority of humans. Everything that had been interpreted in this way must be reinterpreted as the result of minuscule changes in ancestral and other related species, species whose histories simply had to be littered with "precursors of," and "stepping-stones toward," any capacity that had been regarded as uniquely human. There could not be anything you could call a discontinuity. A few holdouts would reluctantly allow a small measure of discontinuity in language, but even here it was widely believed that nonlanguage somehow segued into language, through precursors, across stepping-stones, without any real Rubicon to cross.

Anything else was a no-no—meant giving aid and comfort, even tacit endorsement, to those who were increasingly perceived as enemies, those who still believed humans arose through a unique act of creation. As I have written elsewhere, to suggest that the discontinuity between language and nonlanguage was only part of a much greater discontinuity fell somewhere, on the scale of political correctness, between Holocaust denial and rejecting global warming. Despite the fact that, as an intrepid trio of researchers wrote, "human animals—and no other—build fires and wheels, diagnose each other's illnesses, communicate using symbols, navigate with maps, risk their lives for ideals, collaborate with each other, explain the world in terms of hypothetical causes, punish strangers for breaking rules, imagine possible scenarios, and teach each other all of the above." This, and much more; the list compiled by Derek Penn and his colleagues barely scratches the surface of all the things humans can do that no member of any other species has even come close to doing.

If the gap between humans and other animals is as small as we've been told, what in the world could possibly be this minuscule difference that makes all other animals do so little and us do so much? So far as I'm aware, none of those who argue for continuity between humans and other species have ever realized, let alone admitted, that *each time the gap is minimized, the manifold, manifest abilities of humans become more mysterious than ever.*

Does that mean we must accept some all-powerful deity, or some enigmatic Intelligent Designer?

Of course not. The evidence for evolution is far too widespread, far too strong: somehow, somewhere, perfectly normal evolutionary processes have produced the difference, whatever it is. We've just been lazy. We haven't done our due diligence. And in the interests of dogma, we've kissed objectivity goodbye. Discontinuity exists, and that discontinuity is not limited to language—it extends to all aspects of the human mind. We have, first, to admit that it exists. Then we have to figure out how evolution could have produced it.

In nature, a tiny change can sometimes lead to a phase transition. A few degrees down, liquid water becomes ice. A few degrees up, it becomes steam. Steam and ice and water are things that behave in totally different ways, yet the boundaries between them are, pun intended, still just a matter of degrees.

Or take living creatures—take flight in insects. Nobody's sure how insects developed flight. Did they enlarge the gills they'd used in their previous aquatic existence until they were big enough to glide with? Did they grow vibratory devices for cooling purposes that, one fine day, lifted the first of them into the atmosphere? Whatever happened, those first flights would have been over in seconds, but a barrier had been breached, a totally new realm had been opened up, a realm with new and limitless possibilities. Now there's a discontinuity for you.

What powered the human mind was the intellectual equivalent of flight.

Penn and his coauthors assumed there were two discontinuities, not one: a particular discontinuity in language and a more general discontinuity in cognition. They couldn't see how the first could have caused the second. They didn't show how the second could have caused the first, either. What they failed to face was the profound improbability that, in a single, otherwise unremarkable lineage of terrestrial apes, two evolutionary discontinuities of this magnitude could have emerged.

It doesn't make sense. One would be bad enough. And in this book, for the first time ever, I'm going to show you not just how language evolved, but how language caused the human mind to evolve.

But why did any of this happen?

If prehumans broke from a communicative pattern that had served all other species well for half a billion years, they must have been driven by some kind of need—a very strong need, surely, to produce such radical consequences. Perhaps they developed a new kind of behavior that required them to communicate in ways beyond the range of previous communication systems. But the neo-Darwinian consensus of the twentieth century seemed to rule out any such development.

According to George Williams, an icon of modern evolutionary biology, "Adaptation is always asymmetrical; organisms adapt to their
environment, never vice versa." On the face of things, this sounds in-
disputable; how could the environment—rocks and trees, wind and rain
and sunlight—adapt itself to you and me? But a consequence of
Williams's position, one widely shared among evolutionists, is that evolution becomes a one-way street. "Adaptation" makes it sound as if
organisms are doing something positive, but that's not what it means. It
means that animals, including us, are not agents of their own destiny,
but automatically throw off random genetic recombinations and occasional mutations from which the environment selects. That's natural
selection. Nothing in the animals' actual behavior has any effects or any
significant consequences. This is the view of evolution taken to its logical extreme by Richard Dawkins's "selfish gene, genes are everything"
approach.

Now if what I just described was the whole evolutionary story,
there'd be no point in searching through the course of human evolution
for some special, unique behavior that could have triggered language.
There couldn't be one. Our ancestors must simply have gone on having
sex with one another and recombining their genes and tossing out the
odd mutation until one fine day they hit the jackpot with some combination that made language, at least in a very simple form, possible. And
then once they were capable of language, it was what the French call an
embarras du choix; there were just too many things for which language
would obviously be useful. Hunting, toolmaking, social relations, rituals, gossip, scheming for power, attracting mates, controlling children . . . All these and more have been proposed as the original function
of language. After all, these activities were all carried on by other primates. And since we were primates with primate genes, and since genes
were what determined behavior, there was no point in looking anywhere but at our closest relatives, the great apes (who unlike our immediate ancestors had the advantage of being alive and well and available
for study), if we wanted to know how language began.

Irene Pepperberg, who has shown that at least one species of parrot
has as much language potential as apes, called this the "primate-centric"
approach to language evolution.

Let's look a little more closely at Williams's dictum. "Organisms adapt to their environment." Not the environment, note—their environment. The environment as a whole doesn't select for anything. (The weather in Alaska doesn't bother Hawaiian finches.) A species is only affected by the environment that immediately surrounds it. But that environment is in its turn changed, sometimes drastically, by the species that inhabit it. Goats cause deforestation. Worms enrich the soil. Beavers flood valleys. Seabirds dumped so much guano on the island of Nauru that, now that the Nauruans have sold it all, there's hardly any island left. So the selector in natural selection isn't some generalized, abstract "environment"; it's a part of the environment that has already been worked over by its inhabitants. What living organisms did to that environment will then select for new traits in those organisms that will enable them to modify their environment still further, which in turn . . .

Get the idea? It sets up a constant feedback process.

So evolution is no longer just selfish genes mindlessly replicating themselves. It's a process in which the things animals do guide their own evolution. This happens to be a much more user-friendly view of evolution, but that's not why you should accept it. You should accept it because it's closer to the truth.

It's only in the last few years that this view, one known to biologists as niche construction theory, has developed; it's still hardly known to outsiders. Nobody has yet used it to look at language evolution. I'll explain what niche construction theory says in chapter 5. All we need here is the radically changed picture of human evolution that it gives us. No longer is human evolution, and the complex culture that human evolution produced, a one-of-a-kind anomaly. What drives it can now be seen as a process operating in many other species—possibly most species.

Human culture is just the human niche.

It's the way we adapt our environment to suit ourselves, in the same way that the complex worlds of ant nests or termite mounds are the way ants and termites adapt the environment to suit them. We do it by learning, they do it by instinct; big deal. We can do it by learning only because we have language, which is by now the fruit of instinct just as much as a termite mound is. And language itself is a prize example of niche construction.

What this new theory suggests is that people have been seeking the origin of language in all the wrong places. Previous treatments fall into one of two categories. Either language was some exotic gift that fell from on high for no very clear reason, or it was such a simple and obviously useful thing that any of a dozen factors might equally well have selected for it. We'll meet both kinds of explanation in the pages that follow, and see what's wrong with each of them.

From the perspective of niche construction theory, language could only be the logical result—maybe even the inevitable result—of some very specific choices our ancestors made and some very particular actions they performed. To be more precise, they must have started to do something that no species of even remotely similar brain power had attempted, something that could not be accomplished unless they somehow broke through the limitations that restrict almost all other animal communication systems. And of course, once they broke through, once they established a new kind of system, they would have moved into a new niche—the language niche. No matter how crude or how primitive that first system was, it would be subject to the same feedback loop— behavior to genes, genes to behavior, behavior back to genes again— that all forms of niche construction create. Language would change, grow, and develop until it became the infinitely complex, infinitely subtle medium that we all know and use (and take totally for granted!) today and every day of our lives.

I have two goals in writing this book.

First, I have a burning desire to convince you that language is the key to what it means to be human, and that without understanding how language evolved, we can never hope to explain or understand ourselves. I'm not saying this because the evolution of language happens to be what I've been thinking about for the last couple of decades. It's the other way around. I've been thinking about the evolution of language for the last couple of decades precisely because I'm convinced it's the key to understanding humanity, and for no other reason. I didn't have to do it. I don't need the money, not that there's much money in it. I could have stretched out on a chaise longue by the pool with a pitcher of margaritas and blown the days away. But the desire I have to convince you merely reflects my own passionate need to know, to understand, what humans really are—a need I've had all my life.

Second, I want to dispose of some of the many confounding factors that have bedeviled the study of language evolution, that have made it a chaos of conflicting theories, extravagant claims, and irreconcilable positions. One of these factors I've already noted: the "primate-centric bias" that affects so many in this field, focusing exclusively on our genetic continuity with the great apes and ignoring all the environmental and ecological differences between our ancestors and theirs.

Another factor, closely allied with this one, is the belief that the communication systems of other species make up some kind of hierarchy, like a ladder or a pyramid with language seated firmly on top. It's as if the communication systems of other species were no more than a series of botched attempts at language: they did their best, but weren't quite up to it; only we were smart enough to scale the pinnacle. This is what you might call the "homocentric bias"—folk seldom admit to it, but it's colored all too many theories. Watch out for people who talk about "precursors" of this or that aspect of language, or who seek "stepping-stones to language" in the communication of other species: these are some of the signs of homocentric bias.

In reality, the communication system of any species is designed simply and solely to take care of that species' evolutionary needs. There's no evidence anywhere for a cumulative or "progressive" tendency operating across communication as a whole.

A third factor is assuming that language was originally a target for natural selection. This looks like a no-brainer. Language was what evolved, and evolution proceeds through natural selection, so language had to be selected for, didn't it? The question then becomes simply, what selected for it? Was it hunting, toolmaking, child care, social competitiveness, sexual display? All these and more have been picked by some experts as the pressure. Not surprisingly, there's no good reason to choose any one of these over the others; indeed all of them are seriously flawed in one way or another.

The error here, made even by those who think the earliest language was far simpler than the languages of today, is thinking that language could have been a target at all. How could it be a target, even in its simplest, most basic form, when it couldn't even exist until some of the necessary bits and pieces had been assembled?

Instead of asking how language evolved, we should be asking what caused our ancestors to take the first halting steps away from the kind of communication system all other animals had and have. We should be looking at those ancestors' way of life, what they were trying to do and how they did it, and then asking which of the constraints on animal communication those activities would have forced them to break.

If we can avoid all of these confounding factors, we may be able to get past the two head-butting alternatives in which the language evolution debate is all too often framed:

- "All communication systems are on a continuum."
- "Language is a totally different kind of communication system."

Too often, these contradictory positions are argued on ideological rather than scientific grounds: those who want humans to be just another species take the first position, and those who think humans are something very special take the second. We have to realize that the dichotomy is a false one; the second may be true now, but it certainly wasn't then, whenever "then" was. We have to look, more closely than anyone has yet done, at how our ancestors could have first cracked the mold of animal communication, and how that first breakthrough, in a species not so distant from our own, could have unleashed a cascade of change that would radically alter not just communication, but the very minds that communicated.

It's a long story, a complex story.

But is it the true, the only real story?

I can't guarantee that. Science isn't faith. What seemed certain yesterday can look like nonsense tomorrow, yet become possible again the day after. Not because scientists can't make up their minds, but because new knowledge is constantly coming in, because that knowledge inevitably changes (hopefully improves) our picture of reality, and because, not being faith-based observers, we have to ensure that our theories fit that picture.

What I can guarantee is that, on the basis of what we presently know about humans, evolution, human evolution, biology, and language, what you will read in the following chapters represents the best and best-supported account it's possible to get today. What we know may change, and it may no longer be the best, but our knowledge would have to change a lot before that happened. For what I think will remain true, regardless of new discoveries, is the idea that we must look for the source of language not in the things apes do today, but things our ancestors did that apes didn't do.

But that, as they say, is an empirical question.

You be the judge. If you enjoy this book half as much as I've enjoyed writing it, it will have been more than worthwhile.

I

THE SIZE OF THE PROBLEM

Almost all animate organisms communicate with one another . . . somehow.

Fireflies flash. Frogs croak. Crickets, grasshoppers, and the like rub their legs together, or against their wing cases, producing the kinds of sound known as stridulations. Birds perform songs of varying degrees of complexity. Wolves howl. Dolphins emit sonar signals; they whistle too. Some lizards inflate pouches in their necks, or change color. Gibbons engage in bizarre duets that can last for an hour or more. Apes and monkeys have a range of strategies: hoots, barks, gestures, facial expressions. Bees dance. Ants do it with chemistry. The means different species use for communication are so bewilderingly diverse and so different from one another that you might well think something pretty complex is going on.

It isn't.

A decade ago, Marc Hauser published what is still the most thorough and complete study of animal communication systems (ACSs for short; sorry about that, I loathe acronyms with a passion, but if you had to read "animal communication system" as often as I'll have to write it in the next few chapters, you'd understand—even forgive). He found that all the information conveyed by ACSs falls into three broad categories. There are signals that relate to individual survival, signals that relate to mating and reproduction, and signals that relate to other kinds of interactions among members of the same species—call them social signals. Some signals are hard to fit into a single group. For instance, a signal of appeasement, used in confrontations when your enemy looks to be winning, is on the face of things a social signal, but it could also fall under "survival"—if you don't make it, you could get killed. But no signal falls outside of those three areas. No ACS can be used to talk about the weather, or the scenery, or your neighbor's latest doings, let alone to plan for the future or recall the past.

Of course it would be an enormous benefit for any animal if it could recall the past, noting all the mistakes it had made, and plan for the future, eliminating those mistakes. Such an achievement would maximize an animal's fitness, which is biology-speak for saying the animal would live longer, have more offspring, and spread its genes more widely. And that's what evolution's about: who dies with the most kids wins. So you may well wonder why we alone have language—why we don't inhabit a Dr. Dolittle world, where we could chat with chimps, converse with cats, debate with dogs, rap with rabbits, and yak with yaks, while all these creatures did the same things with one another.

The answer is, evolution doesn't develop things just because they'd be useful for a species to have. Evolution is a minimalist. It doesn't do a lick more than it has to. And it's also limited by what it has to work with. What it has to work with are the bodily shapes and mental abilities that exist in any species at any given moment, and the behaviors that those shapes and abilities make possible. Since within any one species those shapes, abilities, and behaviors can't vary all that much, almost all evolutionary changes are gradual and small. Very occasionally there may be a tipping point, but for the most part, nature doesn't even look like it makes leaps.

So the means by which animals communicate—all the flashes, calls, gestures, and so forth I mentioned at the beginning of this chapter—are seldom if ever things that were designed from the beginning to communicate with. Rather they are modifications or stylizations or amplifications of things animals would do anyway, things that when they started out may have had little or nothing to do with communication. This was the conclusion of the earliest ethologists, scholars like Nikolaas Tinbergen and Konrad Lorenz, and though interpretations of the function and significance of ACS units have changed radically since the 1950s, our understanding of where they came from hasn't changed.

Over time, through frequent co-occurrence, these original behaviors became associated with certain situations, and hence with the kinds of messages appropriate to those situations. Their users didn't consciously mean to communicate in the way we do when we want a window closed and say, "Please close the window": ACSs aren't just a cheap substitute for language, but something entirely different. Their users, in the process of reacting to situations, provided clues as to how other animals should react in those situations; interpreting such clues correctly improved those animals' chances of survival. Thus in a confrontational situation between mammals, shrinking postures and high-pitched sounds indicate an intent to appease the aggressor. Among songbirds belonging to species that defend territory, songs of a certain type and intensity suggest willingness to combat an intruder. And so on.

This already makes language sound unlikely. The simplest and most straightforward source for it would be sounds drawn from the ACS of the last common ancestor of chimpanzees and humans. But if those sounds were anything like the sounds chimps make, the chances of modifying or ritualizing them into words, not to mention sentences, look small. And that's even before you consider problems of meaning.

To lift yourself by your bootstraps, you first have to have bootstraps.

Why other animals have so little to talk about

It gets worse. Why do ACS units contain only survival, mating, and social signals? It's because those areas and only those areas are ones where signals can significantly increase an animal's fitness.

Look at survival calls. These include predator warning calls and food calls. A predator warning call doesn't improve the survival chances of the animal that makes the call. In fact it reduces them; it calls attention to that animal, makes it stand out as a target. But it does improve the survival chances of that animal's close relatives, who carry many of the same genes. This is what biologists mean when they talk about "inclusive fitness." You don't do stuff just to increase your own chance of sending more offspring into the world—the same purpose is served if you increase the chances of your siblings or other close kin doing so.

Once it was thought that things like warning calls were entirely automatic, like the way you blink if someone pokes a finger in your eye. The poor animal saw a leopard and yelped; it just couldn't help it. Now researchers have found that animals, though unable to speak, aren't that dumb. If they're alone, they don't call. If they're not with close relatives, they're less likely to call than if they're with immediate family.

Food calls—signals that announce the discovery of food, sometimes even its kind or quality as well as its location—can again be unhelpful to the individual if it means sharing a tasty tidbit with others instead of hogging it alone. But the same standard of inclusive fitness applies: benefit your brother and you're giving your own genes, at least some of them, a boost. So all survival calls relate directly to increasing fitness.

Now take reproduction signals. These may involve advertisements of immediate availability, such as the swelling of female genitalia in some primates when they are in estrus, or they may merely announce "I am a male/female of species X." At the other extreme, they may involve elaborate courtship dances, or the construction of complex artifacts (such as a decorated bower) to attract females. The simpler signals merely ensure that the right sexes of the right species get together at the right times. Clearly, if at one time there were animals that signaled their sex and their species alongside other animals that didn't, the first lot would meet and mate oftener than the second, so eventually every member of the species would make those signals. The more complex signals indicate not only that a mate is available but suggest that a mate of the highest quality is available. As Darwin long ago pointed out, female choice—the desire of any female to secure the best breeding stock, to bag a mate that will send her genes further into futurity—forms one of evolution's most powerful engines. So again with reproduction we have a set of signals that directly increase fitness.

Finally, let's turn to social signals. These don't have to be social in the sense of friendly; they don't even have to be limited to social species. They can relate to any kind of interaction between members of the same species. Take for example solitary birds that defend a territory, perhaps with a single mate. The signals they send to discourage intruders fall into the social category just as much as more intimate signals like the "nursing poke" used by infant apes to get their mothers to feed them. While the increase in fitness that results from these signals may not always be as direct or as obvious as with survival or mating signals, it is not insignificant. The animal that makes a rival back off without having to fight avoids possible injury or death. The animal that encourages others to groom it benefits from more than the elimination of vermin. Given that the favor is returned, affiliative bonds are then established, status within the group is enhanced, access to sexual or nutritional opportunities is increased. Better living means longer living and more progeny—once more, increased fitness.

Why language is so anomalous

Now that we've established two of the most basic features of ACSs—that they grew from behaviors not originally meant for communication and respond only to situations that directly affect fitness—we can begin to realize the enormous size of the problem that language poses for the biological sciences.

People often think that the core of this problem is the uniqueness of language. It isn't. Lots of things about humans are unique: bipedalism and absence of body hair (among terrestrial mammals, anyway), the precision grip of thumb and forefinger, even the whites of our eyes. Lots of other species have unique features too: the elephant's trunk, the giraffe's neck, the peacock's tail. And the hammering of woodpeckers, the heat-sensing of pit vipers, the trap-digging of ant lions are behaviors as unique as the physical forms of elephants, giraffes, or peacocks. But no other unique feature of any species is as isolated from the rest of evolution as language is.

Bipedalism isn't all that special. Birds managed it. Kangaroos come close. Closely related apes get up and hunker around on their hind legs from time to time. Our grip differs from that of ape fingers only in its greater range and degree of control. Hairlessness is unique, in our case, only because it's lifelong; the young of many mammalian species emerge naked from the womb and only later grow hair.

Instead, let's compare another feature that's not human but is genuinely unique: the elephant's trunk. In his book *The Language Instinct* the psycholinguist Steven Pinker actually uses the elephant's trunk to make language seem less of an anomaly than it really is. He asks his readers to "imagine what might happen if some biologists were elephants"—as with language, some would say the trunk was too unique to have evolved, others that it couldn't really be unique at all. But unique stuff *can* evolve through natural selection, so Pinker insists that "a language instinct unique to modern humans poses no more of a paradox than a trunk unique to elephants."

He's wrong. An elephant's trunk results from hyperdevelopment of the nose and adjacent parts of the face in the common ancestor of elephants and hyraxes, and anatomists can point to the exact physical ingredients that went into its makeup. But Pinker doesn't tell you what ingredients went into the making of language. (And anyway, isn't it a bit weird to compare a behavior with a body part?)

Uniqueness isn't the issue. Unlikeness is the issue. And that is something Pinker, like everyone else who writes on language evolution, doesn't really tackle. For every other "unique" thing that's evolved, you can see what was there before it, what evolution had to work on in order to produce it. Not with language.

Take what looks on the face of things the best, if not the only candidate: the communication system of the last common ancestor. But to get from *any* ACS to language would involve two tasks, just for starters. First, evolution would have to find the raw material—some already existing behavior that could be taken and twisted and refined into an appropriate medium. Second, and this is a task orders of magnitude harder, it would have to uncouple this new system from currently occurring situations involving fitness.

That's actually three tasks in one. The system would have to be uncoupled from situations, from current occurrence, and from fitness. Let me explain.

ACS units—all the calls, flashes, and gestures that constitute ACSs—are all anchored to particular situations: aggressive confrontation, search for a sex partner, appearance of a predator, discovery of food, and so forth. They would be meaningless if used outside those situations. Language units—words, manual signs—are not. They're meaningful in any situation. If I say, "Look out, a tiger is about to jump on you," you may know I'm just kidding, but you know perfectly well what the words mean—they mean exactly what they'd mean if a tiger really was about to jump on you.

Some linguists and philosophers may still tell you that words relate directly to individual objects in the world—dogs, chairs, trees—but they don't even do that, or rather they do so only indirectly, via the concepts of these objects that we have in our minds. If I say, "Dogs bark," what actual dogs am I referring to? Big dogs? Brown dogs? The dogs down the road? Obviously not. All dogs, then? Not necessarily. I didn't say all—my statement cannot be refuted by a barkless dog. What it means is, "Dogs as a general rule bark," or "Barking is a fairly reliable sign of dogginess." Well, just point out to me "dogginess" or "dogs as a general rule." You can't; there's no such critters. We have what may be vague but are fully functional ideas of what dogs are like, and that's what we're referring to. If we want to refer to a specific dog or dogs we can't just say "dog" or "dogs"; we have to say "this dog," "those dogs over there," "the dog with the waggly tail." So in order to get to language, the reference of meaningful units—signs or words—has somehow to be shifted from concrete situations to the concepts we have of particular things in the world.

But what ACSs are grounded in aren't just any old situations. They're situations that are occurring right now, at the very moment the ACS signal is being waved or flashed or yelped. No animal can use a predator alarm call to remind its fellows about the predator that appeared yesterday, or the predator that often hangs around the water hole. No chance of an advance warning, no reprise of what went wrong last time. Each utterance of an ACS unit is tied to whatever is going on in the immediate vicinity right at that moment. Words, on the contrary, are relatively seldom used about what's going on before our eyes. We can usually see that for ourselves, so what would be the point? We still have body language; for things like showing how far we'll push a confrontation, or how strong our sex desire is, good old body language works as well for us as for any other species, often much better than words. On the other hand, with words we can do stuff way beyond the here and now. We can exchange ideas about things infinitely remote in space and time, things we may never have seen, even things like ghosts or angels that may not exist. So somehow communication has to be released from bondage to what's happening right now.

Finally, there's freedom from fitness. We have seen how the function of ACS units is to improve fitness; no unit even comes into existence unless it improves fitness in some way. Some people have conjectured that language as a whole increases fitness. Now it may well be that at some stage in evolution, ancestors of ours who had more developed language skills left more offspring than those whose skills were less developed. But though this is a plausible conjecture, there's zero evidence for it, and in any case it's a totally different issue. The point is that no ACS signal occurs in any situation that doesn't directly involve fitness. And this certainly isn't true of words or signs. They can refer to anything at all, whether it has any connection with fitness or none. And, leaving aside one or two exceptions like "Fire!" or "Help!," a word can't, in itself, by itself, contribute in any way to fitness. And these exceptions, when you come to think about them, are more like ACS calls than regular words—they're tied to situations in just the way ACS signals are. If you doubt me, try shouting "Fire!" in a crowded theater, or ask yourself whether "fire" works the same way in "Help! Fire!" as it does in "There's nothing like a nice warm fire on a winter's evening."

Let's do another thought experiment. There must have been a time when the first system that broke the ACS mold—the first protolanguage, let's call it—had ten units or fewer. So think of any ten words or signs that, singly or in combination, would increase the survival chances and/or procreative capacities of their user.

There are some constraints on this exercise, of course. There's no point in saying things that could equally well be conveyed without words. Expressions like "I'm hot!" or "Check this for size!" don't qualify; nonlinguistic means can express them more than adequately. Also, first words have to be plausible as first words; they can't be abstract, but must be things whose meaning could easily be demonstrated, by mimicry, pointing, or whatever. Finally, their message can't depend on the way they're assembled; most people in the field agree that words came before syntax. So while you're allowed to string words together in an ad hoc fashion, the final meaning can't depend on the positions the words hold with respect to one another.

No prizes, I'm afraid. If I were to offer prizes, you'd have to swear you hadn't read past chapter 5, and I'd have to believe you.

Why is this experiment important? Why "ten words or fewer"? Why not "twenty" or "fifty" or "a hundred"? I mean, give language a break; what use would you expect fewer than ten words to have?

Well, the point is that if the first few words did not have some immediate and tangible payoff that couldn't have been obtained by simpler means, language would never have gotten past ten words, or even that far. Evolution has no foresight. It doesn't think, well, if we can crank language up to say fifty or maybe a hundred words, here's all the nifty things we can do with it. Actually I'm being generous with ten. From word one, language had to pull its adaptive weight, confer some kind of benefit. If not, then nobody would have bothered to invent any more words.

ISSUES OF NEED AND UTILITY

So to obtain freedom from fitness, freedom from situations, freedom from the here and now all in one package, so to speak, represents an enormous task, a task without precedent in the three billion years since the first primitive life-forms emerged on this planet.

Think of it. Think of all the billions of species there have been, in all those years. Not a single one of them failed to get by with a standard ACS. All the things they needed to do could be done with that ACS. And in ACSs in general, there was nothing you could call progress.

You might think that, since chimps are more complex than dogs, and dogs are more complex than crickets, chimps would have more complex ACSs than dogs, and dogs would have more complex ACSs than crickets. It's true that there's a rough—a very rough—correlation between the complexity of a species and the number of units its ACS has. Fish tend to have more than insects, mammals more than fish, primates more than other kinds of mammal. But that's on average: ranges overlap, and the systems themselves, for all the very different means they exploit, are all startlingly alike. They all share the same limitations: they all consist of single, unrelated signals that can't join with one another to make more complex messages, can't be used outside of particular situations, can't do anything but react to some aspect of the here and now.

If all species but ours get along with such systems, there can be only one reason. Other animals didn't get language because, bottom line, they didn't need language.

I can hear people saying, "No, no! Their brains weren't big enough!" I'll deal with big brains in a minute. For the moment, let me point out that the following species have been proven experimentally to be capable of learning very rudimentary forms of language: chimpanzees, gorillas, bonobos, orangutans, bottlenose dolphins, African gray parrots, sea lions—all of the species most closely related to us and some that are more distantly related. And that's pretty much all the species people have tried to teach language to. I don't know of any cases in which people have tried and failed, though I wouldn't bet on your chances with frogs. Seriously, it looks like any species with a sufficiently complex brain ("sufficiently" still being a black box) can acquire some kind of protolanguage, so it's need, not brain size, that's the most crucial factor.

It follows as night follows day that if humans got language, they can only have got it because they had some pressing need for it. Some need, moreover, that no other species (or at least no other species of even remotely comparable complexity) ever had. There must have been something they needed to do in order to survive that they couldn't do within the limits of a standard ACS.

People have always wondered how language began. Only since Darwin have they rephrased the question as "How did language evolve?" But even after Darwin the idea lingered—seldom explicitly stated, but almost always implicit—that we could have gotten into language through things we already did, simply because we could do them better with language. People seem to have thought: Here's all these other animals, all trying to communicate as best they can, then here's us with our bigger brains doing it better—end of story. Hardly anyone seems to have realized the immense uniformity of ACSs, under their superficially different guises, or how closely connected ACSs are to specific requirements of situation and fitness, and hence how radical a departure language was.

What we are being asked to believe is the following:

Every other species took countless millions of years to develop rudimentary communication systems tied inexorably to the things they needed to do in order to survive.

Our species in a tiny fraction of that time developed a vastly more complex system just so that we could do things we already did, and that other species did too, better than those other species could do them.

Put in such stark terms, no one who believes in evolution is going to buy this. Such beliefs survive precisely because they are seldom made as explicit as this. Yet they underlie probably a large majority of explanations as to why and how language began.

For instance, a generation ago the belief was widespread that language had to do with tools—with the making of tools, or perhaps the teaching of others how to make and use tools. Then we found that chimps make and use tools: sponges of leaves to soak water from hollows, trimmed sticks to fish for termites in termite mounds. Christopher Boesch showed that chimps on the Ivory Coast not only used tools to break up palm nuts but showed their children how to do it. Admittedly, their tools were pretty simple, but so were those of our ancestors more than two million years ago. If apes can do this much without language, why would we have needed such a revolutionary step to do similar things?

Then there were those who claimed hunting was the decisive pressure. This idea really never had the plausibility even of the tools idea. First, there's no evidence that our early ancestors hunted, except sporadically, when the opportunity arose, and even then they didn't have the weaponry to deal with anything much above rabbit size. All those fanciful pictures you see of shaggy fellows poking spears into mastodons relate to relatively recent history, probably to our own species, which is less than 200,000 years old (the last common ancestor was a minimum of five million years ago—and quite likely six or seven million). Second, lots of other species (wolves, jackals, lions) hunt cooperatively, and get on fine without a word between them. The clincher came when chimpanzees were observed hunting colobus monkeys. It was as if they were saying, "Hey, you go this way, I'll go that, Bill can block him over there while Fred cuts him off at that branch." But they weren't. They weren't saying a word, but the monkey got caught and eaten just as efficiently as if they had been.

By the 1990s tools and hunting had fallen into disrepute. Now all the talk was of social intelligence. Ethological studies over the previous two decades had shown that the social intelligence of primates, particularly of our nearest relatives, the great apes, was pretty high. Apes formed alliances, played politics. They schemed with and against one another to get access to the most desirable females. They engaged in what researchers Richard Byrne and Andrew Whiten called "Machiavellian strategies," faking one another out, making phony alarm calls—in effect, lying even without words—as they struggled to enhance their status within their group. Indeed, their social life wasn't all that different from the social life of humans. So something involving social interaction must have been the pressure that selected for language.

Here we encounter the primate-centric bias I mentioned in the introduction. What researchers who took the social-intelligence route were doing was looking at apes, seeing what their strongest suit was, and then assuming that human ancestors just took that suit a step further. They seemed not to see that the argument could just as easily be stood on its head. If apes were already so good at handling social relations, how could a handful of words or signs have improved their capacity? Given that human societies are more complex than ape societies, was it a case of "Societies got so complex that we had to have language in order to cope with those complexities," as these researchers suggested? Or was it rather a case of "We got language and that was what made our societies more complex than theirs"? The second is at least as likely to be true as the first.

And in any case, we have to face the same problem we faced with tools and hunting. We're being asked to believe that language, something radically opposed to existing means of communication, came along to help our ancestors do something they already did.

Variants on the "social intelligence" theme quickly diversified, and there's at least one that's coherent enough to merit discussion here.

This is Robin Dunbar's "grooming and gossip" theory. Grooming, which of course includes delousing, is not just a hygienic but also a highly social activity among primates. It's what bonds primates together, enabling primate societies to run (relatively!) smoothly. But it takes time. And if the social group gets too large, grooming, necessarily a one-on-one procedure, takes too long. You don't have time to groom everyone you need to groom and still find enough to eat. So Dunbar suggests that language evolved as a grooming substitute; you can only physically groom one person at a time, but you can verbally groom three or four at a time. And, as Dunbar points out, a lot of our day-to-day chatter consists of a kind of verbal grooming—"buttering people up," as they say.

Why couldn't grooming have consisted of pleasant noises, without meaning—music, in other words? Because, to be efficient, verbal grooming had to hold interest, and what's more interesting than gossip about other group members? Dunbar's students went out to study social conversation and found indeed that most of it was personal gossip. Gossip as grooming, he reasoned, was therefore the original as well as the commonest modern function of language.

Dunbar's theory sounds intriguing and initially plausible. It also escapes the need condition that trips up so many other proposals. If he's right, and group size did increase, then protohumans were faced with a new problem that could well have required an equally novel solution. But did group size really increase among our ancestors? We don't know (yet, anyway—someone somewhere should be working on it). We don't even know, in fission-fusion societies such as chimps have and our ancestors probably had—fluid clusters that are constantly splitting and regrouping—quite what group size means or how it should be measured. And there are some other problems with the theory that we'll touch on later.

For now, all we need to note is that it fails the ten-word test, what you might call the test of immediate utility. There just aren't that many items of gossip you could pass on with ten words or fewer. And if you use most or all of your words for one particularly juicy item, say "Ug seduced your favorite female last night" (assuming, probably counter to fact, that a sentence with even this small degree of complexity could be produced and understood) what will you do for an encore? Keep repeating it? Novelty is the soul of gossip. But there's no way in which a tiny number of words can be permuted to express a wide range of new items. You'd need at least several dozen, more likely a few hundred words before you could begin to do that. But you'd never get that far unless the first few words had already had a substantial payoff.

TESTS THAT THE THEORY MUST PASS

The test of immediate utility isn't the only condition that an adequate theory of language origins has to meet. There are at least four others, and this seems as good a time as any to say what they are:

- Uniqueness
- Ecology
- Credibility
- Selfishness

Let's look at each one in turn.

Uniqueness is there because any serious theory of how language began has to explain not only why language was acquired by humans, but why it *wasn't* acquired by any other species. Even that isn't quite enough. It should explain why, while language has developed to an extraordinary degree of complexity in humans, there isn't a smidgen, not the least sign of it starting to develop, in any other species. Surely a unique effect requires a unique cause. But if the proposed trigger for language is anything that affects other species, it's not likely to be the right one.

This immediately knocks out some otherwise promising suggestions.

For instance, it's been suggested, by Geoffrey Miller among others, that the selective force driving language was female choice—a tried and trusted pressure, given the Darwinian seal of authenticity by the master himself, and confirmed in modern times by both observation and experiment. It explains, for example, why peacocks have such enormous, entirely dysfunctional tails. Females like them. They figure, "If he can survive with a tail that size, he must be hot stuff." And, sure enough, if you trim a peacock's tail he doesn't score so often.

Does that work for humans? Well, there's evidence on both sides. On the one hand, as songwriters Johnny Burke and Jimmy Van Heusen pointed out, "You don't have to know the language" provided that the moon is shining and the girl has that certian expression in her eyes. On the other, there's John Wilkes, eighteenth-century radical activist and notorious debauchee, whose face was ravaged by smallpox. "You're so ugly," a friend remarked, "how do you get all these women?" "Give me but half an hour with one," Wilkes replied, "and I will talk away my face."

But no matter who's right (and as I wickedly pointed out in an article once, if female choice really worked with language, the captain of high school debating should score more chicks than the captain of the football team), the answer's neither here nor there. The eloquence of a Wilkes and the mumblings of a protolinguistic protohuman are apples and oranges. Nobody doubts that, once language had really and truly gotten off the ground, it could, sometimes at least, enhance the evolutionary fitness of its most skilled users. The same is true for almost any language-origins proposal, the "language is power" one for instance: leaders often owe their positions to the gift of gab, and since, as Henry Kissinger told us, powerful equals sexy, they tend to get multiple mates too.

The trouble with all these explanations is, they involve things that a

wide range of species share, certainly those nearest to us. In any number of species, females determine who they'll mate with and pick those
they think best. In many, probably most, primate species, individuals
seek to enhance their status and scheme to obtain more power over
other members of the group. If these factors operate in so many other
species, why didn't they lead to language in those species, too?

Moreover, none of these factors could work unless it first has something to work with. All of them, female choice, power-seeking, and the
like, would surely have driven language once language had started. But
how could they have started it? Female choice has to have something to
choose from—in this case, presumably a whole range of different skills
in using language. Seekers of power and status had to have something
to seek them with; they would have to work at the high end of a range
of language skills broad enough to have a high end. So none of these
things could have had anything to do with the actual birth of language.

The second condition—ecology—merely means that explanations
of language origins mustn't conflict with what we know or can deduce
about the ecology of our ancestors. That includes evidence drawn from
the fossil and archaeological records, which are admittedly scanty and
can seem contradictory at times. But this is no excuse for violating the
ecology condition.

One thing that amazes me about the language evolution field is how
often people ignore this condition. Primate-centrists are the worst cul-
prits. Since apes make such convenient and accessible models, and
since we share so much of their DNA, primate-centrists assume protohumans must have behaved pretty much as modern apes do. If there
seem to be big differences now, well, recent civilization has just made us
don disguises and draw veils over our basic ape nature.

As we'll see in chapter 6, that's a long way from the truth. Our
remote ancestors may not have been much smarter than their ape
cousins, but they lived in dramatically different environments and made
their living in completely different ways. Unless you believe in universally uniform genes that impose identical behaviors wherever they
occur—something modern biology has decisively refuted—you have to
realize that partially arboreal, forest-dwelling apes make pretty poor
models for the ways in which protohumans behaved.

As for the third condition, credibility . . .

It's London, spring of 1998, the Second International Conference on the Evolution of Language. First thing to hit me was the cropped cannonball head and unapologetic London accent of the sociologist Chris Knight, asking without any preamble whatsoever:

"What does your theory have to say about the problem of cheap signals?"

"Er . . . duh," I riposted eloquently.

I was coldcocked *and* blindsided. Cheap signals? What the hell were they? But Chris knows what he's talking about, so I boned up fast, and this is the problem.

In the 1970s game theory began to be applied to biology. Wouldn't it be the case that in a population where each individual strove for its own genetic success, there'd be payoffs for cheats and deceivers? Animals that exaggerated their prowess as potential mates might thereby gain access to breeding opportunities they'd never get through honesty. How could any animal know that the signals she was receiving really meant what they seemed to mean?

The Israeli biologist Amotz Zahavi came up with an answer. The harder a signal is to fake, the more likely it is genuine. Anyone can perform a fancy strut for a while, but to permanently carry around a peacock's tail or the spread of a stag's antlers suggests that their owner really is strong enough to sire sturdy children. In other words, signals had to be costly to be credible.

People like Chris were quick to pick up the implications of this for language. Words are notoriously easy to produce. Everyday speech is full of remarks to this effect: "Talk is cheap"; "He talks the talk, but does he walk the walk?"; "Sticks and stones may break my bones, but words will never hurt me." Words are cheap tokens, so why should anyone believe them? This thought could not help but resonate at a time when, with Byrne and Whiten's "Machiavellian strategies" center stage, everyone knew that primates in particular were adept at deceiving one another, even before words came along. But if no one would believe words, whence came the impetus to make up first tens, then hundreds, then thousands of the things?

Like the utility condition, the credibility condition strikes hardest at the very earliest stage of language, and suggests that, at this stage, language could never have gotten off the ground if the content of its first words couldn't be verified immediately and beyond doubt. This, among other things, puts another nail in the coffin of the grooming and gossip theory—even today, we don't believe half the gossip we hear.

Finally, selfishness. Over the last half of the last century, biologists shifted from believing that, at least sometimes, organisms did things for "the good of the species" or "the good of the group" to believing that everything every organism did was for itself, or at best for the genes it shared with close kin. The first point of view, known as "group selection-ism," quickly became a scientific no-no, inviting obloquy and scorn in equal proportions, though nowadays it's slowly beginning to inch its way back in. (It's fascinating to watch these cyclic to-and-fros in science, like the rise and fall of hemlines, but more stimulating, intellectually at least.)

But it's too early yet to junk selfish genes. What look like behaviors "for the good of the species" may well turn out to be purely self-serving behaviors that happen, quite incidentally, to benefit the species as a whole. Dubious though it may be when pushed too far, the selfish gene notion has lit up too many areas of behavior to be lightly tossed aside.

So consider any linguistic act from this perspective. A gives informa-tion to B. Before the act, that information belonged to A exclusively. A could have exploited it for A's own benefit. Now that's not possible. B can exploit it too. What does A get out of this? If the answer is "noth-ing," or even a negative payoff—A has to split a favorite tidbit with B—why should A transmit the information in the first place? If the answer is "B will repay the favor," what guarantee does A have that B will recip-rocate, won't prove a cheat?

In other words, the first linguistic acts, whatever they were, must have been such that the speaker derived (at least!) as much benefit from them as the hearer did.

THE BIG-BRAIN FALLACY

Having reviewed the four tests—uniqueness, ecology, credibility, and selfishness—that any candidate theory of language evolution must pass, let's dispose of the theory that as brains got bigger, our ancestors got cleverer and cleverer until they finally became clever enough to invent language.

This, in some form or other, is widely believed by many highly qualified scientists. For instance, in a recent interview, Nina Jablonski, who is not only head of anthropology at Penn State, but also, according to *The New York Times*, "a primatologist, an evolutionary biologist and a paleontologist," explained that "in order to survive in the equatorial sun, [early humans] needed to cool their brains. Early humans evolved an increased number of sweat glands for that purpose, which in turn permitted their brain size to expand. As soon as we developed larger brains, our planning capacity increased, and this allowed people to disperse out of Africa."

What's wrong with this eminently plausible-sounding story? Lots of things. One, plenty of animals survive in the equatorial sun without increasing the number of their sweat glands. Two, permitting brain size to expand is not the same as obliging brain size to expand. Jablonski makes it sound as if brains were just bursting to expand but were held down by dumb stuff like inability to sweat properly. That's far from true. Brains are highly expensive in terms of energy; animals have brains just big enough to do what they have to do, and anything over that is dysfunctional. Three, no one, to the best of my knowledge, has ever shown a correlation between brain size and planning capacity, for any species, least of all for species ancestral to humans, about whose planning capacities we know zero. Four, you don't need any special planning capacity to get from one continent to another. All you need is a land bridge and feet—thousands of species have done it, most notably the placental predators of North America who, as soon as Central America was in place, exterminated the indigenous marsupial species of South America.

A dog has a bigger brain than a frog, and a dog can do lots more things than a frog can. This can only be because its bigger brain makes it smarter, you might suppose. However, a quarter century ago, the Scottish psychologist Euan Macphail wrote a paper that no one could challenge but everyone could and did ignore, which claimed that when you considered not the range of things animals could do but the actual mental apparatus with which they did them, there were only three levels of intelligence. There were organisms that could associate a stimulus with a response. There were organisms that could in addition associate a stimulus with another stimulus; all vertebrates and even some invertebrates fell into this class. And there were humans, who happened to have language. Macphail didn't know how language made us more intelligent, but you will, if you read to the end of this book.

What is intelligence, anyway? Before we could compare the intelligence of one species with another, we'd have to have a valid definition and a valid measure that, unlike IQ, would work across species. Nobody has yet produced one. So even if there were more differences between the intelligence of species than Macphail claimed, there's no way we could show by any nonsubjective means that one species was smarter or more stupid than another species.

If, as some have claimed, language was an invention of folk with big brains, it would be doubly unique. In addition to being the only system of its kind in nature, it would be the only biologically based behavior that had ever been consciously and deliberately created. And if you believe we can deliberately create biologically based behaviors, I have a couple of bridges you might contemplate purchasing.

But the real clincher is this. Brains don't grow by themselves, of their own volition; they grow because animals need more brain cells and connections to more effectively carry out any new things they are beginning to do. In other words, brain-size increase doesn't drive innovation—innovation drives brain-size increase, and in chapter 5 I'll show you in detail how the process works, courtesy of the new and exciting theory called niche construction (it's changed my whole way of looking at evolution, as I hope it will change yours).

It follows that we didn't get a bigger and better brain that then gave us language; we got language and that gave us a bigger and better brain.

So how could language have evolved?

By this time you may be thinking, Well, how could language have evolved, anyway? How could anything have met all the conditions I've described to you? You may even be thinking, Hey, maybe it didn't evolve. Maybe Intelligent Designers are right after all; maybe it was a magical gift from above, sprung ready-made from Jove's brow, inexplicable (as some have suggested) by any human brain. Or maybe we're in something like *The Matrix*, and the whole thing is a colossal illusion—we don't really have language, we just think we do.

Stay calm, folks. We have language and sure enough it evolved, despite all the seemingly insuperable obstacles that lay in its path.

But I've made one thing pretty plain, you may think. It can't have evolved, as most biologists would claim, from some prior means of communication, some ACS of the last common ancestor that . . . somehow . . . gradually . . . got modified . . . or something. It must have evolved from . . . well, from something else. What, exactly? Well . . . it's hard to say . . . but *something*.

That's exactly what I thought fifteen, twenty years ago. And until relatively recently, for that matter. After all, I was the guy who produced the continuity paradox: "Language must have evolved out of some prior system, and yet there does not seem to be any such system out of which it could have evolved."

Then how did it evolve? In my earlier work I talked about mental representation systems—maps of the outside world and everything in it—that grew in the brain, over countless millions of years and thousands of species, till they got detailed enough to divide up the world into word-sized bits, just waiting to be given their lexical labels. Once these bits—prelinguistic concepts—were ready, then in some rather ill-defined way, connected somehow with protohuman foraging strategies, a protolanguage, quite different and separate from the protohuman ACS, just somehow popped out. After which, a handy mutation turned protolanguage into language.

Blame it on the rashness of youth (after all, I was only sixty-four at the time). And it wasn't that bad for a first try. *Language and Species* was the first book I know of that tried to work through the whole process of language evolution with some degree of detail and depth. The problem was I didn't have a good framework. Niche construction theory hadn't been developed. When I didn't know something, I filled the gap with what the philosopher Daniel Dennett calls "figment" (as in "figment of the imagination"). And I didn't do what I'm doing here: working through the precise relationship between human and nonhuman communication systems in what I'm sure is excruciating detail (sorry about that, but any serious study is like athletic training—no pain, no gain).

Reactions to what I had written seemed only to confirm my position, at least as far as the continuity paradox was concerned (biologists wouldn't swallow the mutation, and they were right not to, of course). I hadn't expected that I'd stop people from believing in continuity, but what did surprise me was that they went on believing without even trying to refute my arguments; blind faith is far commoner in science than we like to admit. So I was not, in any sense of the word, converted by continuists. I converted myself.

It came through trying to think like a biologist. This is not all that easy for people from other fields. What makes interdisciplinary work so hard is that any academic discipline acts like a straitjacket, forcing you to look only in certain directions, blocking other perspectives from view. It takes a good deal of conscious effort, plus a lot of soaking yourself in other people's literature, to overcome this state of affairs.

The process was speeded up by my encounter with niche construction theory, which made sense of a lot of things that had baffled me. I started to rethink the continuity paradox. Suppose, just suppose, that one took an engineering perspective and started asking whether there was anything in an ACS you might plausibly change that would make it more languagelike. If there was, the next question was whether that thing might plausibly arise through constructing a particular kind of niche. If it could, the next question was: Was there such a niche in human prehistory?

The rest of this book is about the answers to these questions.

2

THINKING LIKE ENGINEERS

SETTING THE BAR

Let's pretend we're language engineers, given the task of providing an alingual species with language.

We have to work on a species that has only a standard, average primate ACS. We have to get that species not to full language—that will have to come a lot later—but to something that points away from an ACS, in a direction that could plausibly lead to language. It doesn't have to be a giant step. Better by far if it's a small step, because the smaller the step that has to be made, the more plausible it will be from the standpoint of evolution.

But before we can make a start, we have to know where we're headed. We have to look at language and what it does that an ACS can't do.

A lot of people have tried to do this, but they've set the bar too high.

They compare an ACS with the kind of language that all of us speak today. And they point to evidence like the fact that language consists of three quite clearly distinct levels. Autonomous levels, they call them, which simply means that although all the levels interact whenever we speak, each abides by its own set of rules, each set being different from the other two sets.

There's a level of meaningless sounds: phonology. Not one of the sounds we use in language means anything, in and of itself. But they're not meaningless in the way grunts or coughs or sneezes are meaningless. Put a grunt and a cough and a sneeze together, and what have you got? A head cold? Well, nothing that means anything. Put two or three speech sounds together, and what you've got might be a word. Potentially, at least—to find out whether it is or not, you've got to go up to level 2.

That's the level of meaningful sound sequences: morphology. That means words, and those bits of things we tack onto words, all the *uns* and *dises* and *ings* and *eds* that are meaningful too, but only when attached to a word stem. Now we can provide names for things, or to be more precise, names for classes of things—"dog" doesn't mean this dog or those dogs, it means just a particular kind of animal. Except for one-word exclamations—"Help!," "Fire!," and the like—we still really can't say anything that means very much. For that, we've got to go up again, to level 3.

That's the level of meaningful utterances: syntax. You may mean, but you can't mean much until you start to put words together into phrases, clauses, sentences. But once you can do a sentence, you're home free. There's literally no limit to what you can produce— paragraphs, pages, essays, books, encyclopedias . . . Given the rules of syntax, you can go on churning out language from here to kingdom come.

Now when you see this degree of complexity (and I've only skimmed the surface; each level has its own awe-inspiring convolutions) your only rational reaction is to throw up your hands and say, like the Maine farmer:

"You can't get there from here!"

Well, that was funny the first time around, but the Maine farmer was full of his own cows' product. You can get from anywhere to anywhere, with the right map. And the right map for this particular bit of country is only now going to be unfolded.

If you insist on comparing ACSs with language as it is today, you're just setting yourself up for a fall. There's a much better model to hand.

PIDGINS FLY TO THE RESCUE

I had the great good luck to come to language evolution from the study of pidgins and creoles.

A pidgin is what people produce when they have to talk to other people but don't have a common language. If you want to know more, read my book *Bastard Tongues*. For now, it's enough to know that you yourself may well have pioneered an incipient pidgin, if on vacation in a place where you didn't speak the language you've struggled to communicate your needs to people and they've struggled back to try to make you understand them. The only reason your makeshift efforts didn't evolve into a full-fledged pidgin is that it was just you and a couple of other people over a period of days. If it had involved most people in the same community over a period of years, those efforts would have resulted in a true pidgin, as sure as puppies turn into dogs.

Think back to what you did. You used any words of the other person's language you happened to know, but you didn't put them together in any systematic way. Why not? You might well say, "I didn't know how, in that language." Sure, but what was to stop you from putting them together the way you did in your own language? Partly the fact that they were, in every sense, "foreign words"—they were foreign to you, you had to grope for them, so they popped out one at a time, with big gaps in between while you went looking for the next. Partly because you didn't know all the words you would have needed even for the simplest sentence. You went with what you'd got; when you couldn't think of a word, you used one out of your own language, or some other language you might know, and hoped the other fellow might know or guess what you meant. And if that still didn't work, you pointed or gestured or mimed. You used anything at all that might work.

That's the nearest you or I or anyone will ever come to feeling what things were like at the dawn of language. Still a long way, since it's as hard for us to forget that we already have language as it is for a jury to heed the judge's admonition, "Forget anything you may have read or heard about this case." But it helps. If you've had this particular foreign-travel experience, pause a moment here to savor the memory of it.

Not everyone thinks I'm right about this. Dan Slobin, a psycholinguist (that doesn't mean a linguist who's psycho, it means one who studies the relations between language and human psychology) at the University of California, Berkeley, thinks a pidgin is not necessarily a good model for early stages of language. He points to the fact that people who create a pidgin already have at least one full human language, whereas the protohumans who started language clearly didn't.

Now I have great respect for Dan when he's on his own turf, which is how children acquire their first language. And he's right insofar as the difference he brings up is a real one. But to clinch an argument, it's never enough just to point to a difference. You have to explain why this particular difference *makes* a difference. Most differences don't. There are big birds and small birds, birds that fly and birds that don't, but a bird is a bird and we all know one when we see one.

The same is equally true of any variety of what, for want of a better word, we'll call protolanguage (not to be confused with protolanguages, which are the hypothesized real-language ancestors of existing language families—Indo-European, for example—and seldom more than five thousand years old). Protolanguage is not true language, but it's made up of languagelike elements. Since I introduced the notion in my 1990 book *Language and Species*, it's been accepted by most researchers in the field that the emergence of language as we know it had to be preceded by something intermediate between true language and an ACS, and (by at least some of those researchers) that things similar to this intermediate can still be seen in the world around us—in pidgins, in the speech of the brain-damaged or of infants, or the productions of apes that have been taught various forms of signing.

In determining whether something qualifies as protolanguage, what matters is not whether you (its speaker) do or don't have a language already; what matters is whether or not you're in the same situation, that of having to communicate without a proper language to communicate in. Now the content of protolanguage, what you actually say with it, will vary depending on who's using it—on whether that's a pidgin speaker, or a Broca's aphasic, or a "language-trained" ape, or a child younger than two, or a protohuman at the very dawn of language.

What won't vary at all will be certain limitations—purely formal, structural limitations—on how you can express that content. Regardless of who or what you are, even of what species you belong to, these limitations will reduce you (if you already have a language) or exalt you (if you don't yet have one) to short and shapeless and disconnected utterances.

If you're a human speaking a pidgin, the pieces that the pidgin is built with will be ready-made words from one language or another. If you're a protohuman pioneering language, they won't be. If you're a human speaking a pidgin, bits of the syntax of your own language may pop up here and there, although that's unlikely in the early stages, when you won't be fluent enough even to cannibalize your original language. If you're a protohuman pioneering language, there'll be no bits there to pop up. But in both cases, there won't be anything you could call structure. No third level, because you have no rules, and if you have no rules, you have no syntax. No second level, because although you have words, those words have no internal structure and so can't be broken down into bits like anti-dis-establish-ment-arian-ism ("opposition to the withdrawal of government support from a state-sponsored church") or given inflections to indicate things like tense or number.

Just one level, where what you see is what you get.

But it's still not down to the level of an ACS. Protolanguage and language share one big thing that ACSs uniformly lack.

Combinability.

Only connect (if you've got anything to connect with)

Languages combine lawfully and protolanguages combine lawlessly. In other words, languages have all kinds of constraints on what you may put together with what; protolanguages don't. Where things can be put together, languages have rules about which goes first; for instance, adjective before noun in English, noun before adjective in French. (Yes, I know we say "court-martial," not "martial court," and the French for "good luck" is *bonne chance*, not *chance bonne*, but these are exceptions that go against the grain of the language.) Pidgins and other forms of protolanguage don't have such rules. You can put anything with anything, in any order, provided that the combination is meaningful in some way. But the bottom line is, you can still combine.

ACSs can't. So far as we know, yet. And I would say that no matter how long or how hard we look, we'll never find an ACS that can combine stuff. In a moment we'll see why.

To find animals that can indeed combine communicative stuff is the Holy Grail for those who believe that ACSs segued seamlessly into language—strong continuists, let's call them. Such a discovery would mean that those animals had a true precursor of syntax, and syntax is believed by some to be the only uniquely human part of language.

Consequently, a true precursor of syntax would be a stunning defeat for the "language is something completely different" crowd. Needless to say, any number of ACSs have been searched and researched for the slightest trace of such a precursor.

The latest candidate will give you some idea of how desperate this search is getting.

Diana monkeys and Campbell's monkeys are two species of African monkeys that inhabit the same territory. Both species give alarm calls on the approach of predators, and Diana monkeys respond to Campbell's monkeys' alarm calls just as readily as to those of their own species. With one difference: sometimes a Campbell's monkey will preface its call with what is known as a "boom" vocalization, a brief low-pitched sound that occurs, on average, about thirty seconds before the alarm. Boom-preceded calls usually relate to some fairly distant predator or to some unexplained event that might mean danger of some kind. When Diana monkeys hear a boom-prefaced call, they seldom respond, and usually go on with whatever they're doing.

Klaus Zuberbühler, a researcher at the University of St. Andrews in Scotland who discovered this behavior and tested it experimentally, wisely hedges on whether it really is syntactic. What's puzzling is why anyone would think it might be. In the first place, something that involves two separate species is not very good evidence for what's supposed to be happening in just one of them. In the second, a receptive skill—ability to determine the meaning of a sequence—is no guarantee of an equivalent productive skill—ability to actually combine things. But it's mostly the kind of relationship between the two units, the boom and the alarm, that makes this a dubious precursor for any kind of syntax.

Zuberbühler says that the booms "act as a modifier" to the alarms. Not so; they don't modify them, but cancel them out. Do you know of any language with a word that means "Please disregard the next word"? I don't. Typically in any combination of two language units, be they words, phrases, or clauses, one unit truly "modifies" the other by making its meaning more precise:

The *English* teacher (not just any old teacher).
Shake well *before opening* (if you do it afterward, you'll get it all
    over your shirt).

Male monkeys mate *when they see the typical female swelling* (not just any old time, like us).

This is what I was talking about just now when I said that a combination, whether linguistic or protolinguistic, had to be meaningful in some way. This is the way—taking something (a subject) and saying something about it (a predicate). Predication is one of the most basic and fundamental processes in language. Syntax may not have any animal precursors, but predication surely was the precursor of syntax. If units couldn't first combine on the basis of meaning, they'd never have gotten to where they could combine on a structural basis.

So the next question becomes, if units of language and protolanguage can combine, and units of ACSs can't, why is this? Is it just happenstance? Are the animals not as smart as us? Or is there a principled reason why they can't, one that makes looking for syntactic precursors among animals a waste of time?

ANIMAL WORDS?

The search for syntactic precursors mightn't be such a waste of time if animal calls were, in fact, precursors of words.

That's the other Holy Grail hard-line continuists seek for: things in animal communication that are precursors of true words. The best candidates so far are the alarm calls of monkeys, especially the most thoroughly researched of these: calls made by the vervet monkey of East Africa. Indeed, the poor vervets must by now be heartily sick of being dragged in every time anyone writes about language evolution.

As we saw above, more than one species of monkey give predator alarm calls. The vervets' calls are just more varied than most. They have a call for eagles, a call for leopards, and a call for snakes. Why can't we say that these calls are in effect the vervet "words" for eagle, leopard, snake?

Because, as I pointed out in chapter 1, while any word can be used in the absence of what it refers to, no animal call can. Even if a call is used deceptively, to distract aggressors or secure a tasty tidbit, those who hear the call have to assume that a predator's really there. If they don't, the ruse won't work. We may choose to call this "meaning," but it's quite different from the way words in any human language work. Realizing this, some people have chosen to call it "functional reference." That's a way of saying, well, words refer fully, in the sense that you can use them in meaningful ways regardless of whether what they refer to is there or not. But since the leopard call isn't used for anything other than leopards, it has the effect of drawing attention to leopards, and thus discharges the most basic function of reference, which is to pick something out and direct your attention to it.

However, there's a function more important than reference that these calls perform, and that's eliciting a specific response from the hearer, as follows:

Eagle call: look at the sky, get ready to hide in the bushes.
Leopard call: look around, get near a tree you can climb real quick.
Snake call: look down at the ground all around you.

Do these sound like the names of different animals? If we try to translate them into humanese, the animal's name may not even be included in the translation. Take the eagle call. Does it translate as "Look out, an eagle is coming!" or "Danger from the air!" or "Quick, find the nearest bush and hide in it!"? Any of these three seems more efficient, more *functional*, than simply "eagle."

(Note that even here, the potential ambiguity is not the same as the ambiguity sometimes found in words. Ambiguous words mean totally different things. A "bank" is either a place where money is kept or the side of a river; "rape" is either a violent sexual act or a commercial crop. But translations of animal calls represent various possible interpretations of the same thing. Remember this; it will become important in the next chapter.)

What do my three translations of the eagle call have in common?

They are all complete utterances—complete in themselves.

How does the word "eagle" differ from that?

It isn't complete in itself. It tells us something, but not enough. Is there an eagle here right now, or are you talking about yesterday, or the chance we'll meet one tomorrow? Are you making a general statement about eagles, or a particular statement about one eagle, or just listing major bird species? It could mean any or all or none of the above.

For me to know what you're talking about, you have to predicate. You have to combine "eagle" with some other word or words that will tell me which of the many possible things you mean. But for me to know what an alarm call means, you don't have to predicate. The call is enough. I'm up the tree or into the bushes already.

Now we can see why signals in ACSs are never combined.

It would make no sense to combine them. They're not words that have to be combined to form a particular meaning. They're specific responses to specific situations, complete in themselves, and more than that, they're responses that have had, in the past, a demonstrated capacity to improve the fitness of those that used them. If those responses hadn't produced longer lives and more offspring for their users, evolution would have erased them.

It's not that animals are too dumb to put things together. Just that the calls and signs and all the other things they communicate with weren't designed to be put together. And if you did put them together, one wouldn't "modify" the other; together, they would mean exactly what they meant separately. One wouldn't change or affect the other in any way.

This hasn't always been apparent to everyone. In 1964 the journal *Current Anthropology* published an article called "The Human Revolution," by Charles Hockett, one of the leading linguists back then, and his colleague Robert Ascher; the journal thought so highly of this article that it was reprinted, unchanged, twenty-eight years later. (Until 1990 or thereabouts, the pace of change in language evolution studies was, indeed, glacial.) Hockett's intuition was that language began when some protohuman, encountering a situation in which there was both food and danger, blended the food call with the danger call. Then this, the first combination of meaningful units, led to more of the same, and language was up and running.

Hockett's analysis ignores the following facts:

Words combine as separate units—they never blend. They're atoms, not mudballs.

To a naive animal, a blended call would probably be meaningless.

Even if the blend had been interpreted, how could it have made sense? If it had been a predication, there are only two possibilities:

"Dangerous food"? Unlikely; danger calls, as we've seen, at least roughly specify the source of the danger without further addition. No animal I've ever heard of has an alarm call for poisonous substances.

"Edible danger"? Come on!

All the blended call could have meant was, "There is food, but there is also danger." But, just as I said, this is no more than the sum of what the two calls mean in isolation. As such, it would have moved us not one inch closer to anything you might call language.

The dream of strong continuists is to find precursors of words and precursors of syntax among other species. That would be the easiest and most obvious way to establish true continuity between ACSs and language. But it isn't the right way, simply because while words (or manual signs, or any similar kind of linguistic unit) have little meaning until they're combined with other words, animal calls (or any other ACS units) mean no more when they're combined than they mean in isolation.

So why would any animal in its right mind even want to combine them?

It's a waste of time looking among other species for precursors of words or precursors of syntax, because animal communication was not designed by evolution as an inferior substitute for language. It wasn't that animals were slowly and stumblingly trying to get nearer to language, but didn't quite know how. What we've been looking at as if they were ACS limitations are really only limitations from our own peculiar perspective. For other animals, ACSs do their job just fine. It was only some aberrant ancestor of ours that needed something a little different (and got much, much more than it had bargained for).

So if we want to demonstrate real continuity in evolution, we should be looking not for linguistic precursors, but for some point of flexibility in ACSs, some point where the right selective pressure could force a distortion that might ultimately lead to the creation of words, and, later on, the creation of syntax. Because these things—words and syntax—are total evolutionary novelties, things useless and meaningless outside language. Things whose like had not been produced by evolution in all the three-billion-plus years it had been working—not because, in all those years, evolution had "failed to produce language," but because it had succeeded in producing something wholly different from language. Not some poor limping thing longing to be language, but a tool that served the needs of its users perfectly well.

Talk about primate-centric—people who look for precursors of language are homocentric. Instead of looking at communication objectively, from a neutral perspective, they seem like they're strangled by language—locked into the worldview of one rather peculiar species.

To find where ACSs are flexible, we must yet again compare them with language—not so we can disparage them, but so as to better understand the very different ways in which they work.

One of the things ACSs don't do but language does is refer to anything that isn't right there, at the moment you make the call, immediately within the range of your senses. Once more we must ask the question: Is this accidental or is there a principled reason that things are the way they are, and not otherwise?

Philosophers of language might say, it's because the signs of ACSs are indexical, not symbolic.

An indexical sign is one that points directly at its referent. Vervet predator warnings are good examples of indexical signs. A symbolic sign, however, can stand in place of its referent, even when that referent is thousands of miles away or thousands of years back in history.

But this just names the difference; it doesn't explain it.

We could ask, why are ACS signs indexical, rather than symbolic? But a more revealing question would be, which comes uppermost, the informative or the manipulative?

We must tread carefully here. All communicative acts are informative, in some sense, and in that sense both ACSs and language are both informative and manipulative. Body language—part of the human ACS—is informative; if in spite of your placatory words your body language shows me that you're angry, that's information, a kind of information I wouldn't have had if you hadn't used body language to express your anger. Similarly, any linguistic act can be manipulative; a purely factual statement about the weather could be aimed at convincing you to stay home with me, rather than go out with someone else. So it would be easy, and true, even if quite uninformative, to say: ACSs are both informative and manipulative, and so is language. What's the difference?

The difference is that an ACS is primarily manipulative and only secondarily informative, whereas language is primarily informative and only secondarily manipulative.

ACSs may provide information, but that information is merely a by-product. Their primary function is to get you to do things that enhance my fitness. (If they enhance your fitness too, you're just lucky.) But if ACSs are designed to respond to situations and manipulate other individuals, you can see why they have to be bound to the here and now. You can't respond to a situation that's distant in time or space (at least you couldn't until they invented TV). You can't manipulate somebody who isn't there, or do it at any time other than the immediate present. What looks to us like a limitation is, in ACS terms, simply a logical necessity.

Language, however, puts information first and manipulation second. Suppose I were to explain to you Einstein's theory of relativity, or Chomsky's theory of a biologically based language organ. My purpose in giving you this information might well be that of impressing you or even ultimately mating with you (although anyone who could be induced to mate by such means would have to be pretty weird). But I would be *trying to manipulate you by means of information* rather than just *inadvertently giving you information in the course of manipulating you.*

It follows that language doesn't need to be bound by the here and now. Information (whether used to manipulate or not) can be about things that have already happened or things that might happen but haven't done so yet. It can be about things before your very eyes, but it's much more likely to be about things that aren't, because an important—maybe the most important—feature of information is its novelty. In most contexts, old information is plain boring. (The big exception is bonding, whether of lovers or party members—have you ever heard a politician's speech that contained anything you hadn't heard umpteen times already?) In contrast, ACSs endlessly repeat the same old signs for the same old situations—novelty would be disruptive, dysfunctional. And if the situations weren't ones that repeated endlessly, evolution would never have gotten around to making signs for them.

By now it should be clear why ACS units are indexical and language units are symbolic.

ACS units are indexical because they're designed to manipulate others. Those others have to be right there in the present time at the present place if they're going to be manipulated. So even if information is exchanged, it can only be information about the here and now.

Language units are symbolic because they're designed to convey information. Information can be past, present, or future, here, there, or anywhere. But to the extent—a very considerable extent—that its value lies in its novelty, it had better not be about the here and now.

But this, of course, is no help at all in explaining how anything could have come to be symbolic in the first place.

WHICH WAY TO THE RUBICON?

A decade ago, Terrence Deacon (then at Boston University, now at UC Berkeley) published a widely acclaimed book called *The Symbolic Species*. In it he claimed that what most sharply distinguished humans from other species was the capacity to create and use symbols. In reviewing it, I said I thought he was wrong and that the really distinctive thing was syntax. And later on, we debated the whole thing publicly, twice (in Seattle and in Eugene, Oregon). But quite recently I came to the conclusion he was right and I was wrong—about symbolism versus syntax, at least.

Certainly, the reason I criticized his book was not the right one. The right one, as I now see with the twenty-twenty vision of hindsight, was that he didn't deliver on his implied promise. There are chapters and chapters on why animals don't get symbols and why we must have gotten them in order to have become the kind of creatures we are. But there's nothing about how we got symbolic words. About how we got symbolism, sure: it was through ritual. And which ritual, in particular? Would you believe marriage?!

No, to be fair to Terry, he wasn't suggesting that the first words were "Do you take this woman . . ." He had in fact what was a rather good argument, at least from an anthropological perspective. It was that among prehumans, where men went off to get meat and women stayed close to home collecting the veggies, there was always the chance some guy would sneak back and mate with your mate. Since the meat you brought back was to share with her and her children, you ran a substantial risk that all your efforts would go to promote your cuckolder's genes at the expense of your own. So, to avoid all the stresses and strains, jealousies and conflicts, that would result, some form of generally accepted pair-bonding ceremony had to evolve. And indeed, marriage of some sort seems to be universal in human societies (though I doubt it reduces promiscuity all that much).

But the nearest Terry got to explaining how symbolism went from (admittedly wordless) rituals to actual words was to claim that although "vocalizations" were found along with all these "ritual gestures, activities and objects," "probably not until *Homo erectus* were the equivalents of words available." How did they become "available"? How did anything get to *mean* anything? Not a word.

Yet I'm convinced now that Terry was right in his main contention, that symbolism was the Rubicon that had to be crossed for our ancestors to start becoming human. I had argued for syntax because, while trained apes could be taught things that in meaning were roughly equivalent to words, and while they could (apparently without much if anything in the way of further instruction) string these words together into a kind of protolanguage, they'd never acquired anything you could call syntax, even though, in at least one experiment, simple elements of syntax had been explicitly taught to them. But syntax, I began to realize, may have become possible only because two million years of protolanguage use brought about significant changes in its user's brain. If that was so, it was ridiculous to treat something apes never had a chance to get to as the main distinction between them and us. It made far more sense, as Deacon had proposed, to see the main distinction as arising at the very beginning of language—at the first step, the creation of symbols, that set the whole process in motion.

For if it's useless to look for precursors of words, or precursors of syntax, there's nothing left to do but look at the units found in ACSs and see if there are any, anywhere, that might, under special circumstances, take on at least one of the properties that symbolic units—words, or the signs of manual sign languages—possess.

And as we have seen, the most salient characteristic of symbols is that they can refer to things outside of the here and now. This capacity is something linguists generally refer to as "displacement."

So let's review Marc Hauser's tripartite division of ACS units into social signals, mating signals, and survival signals. Which of these classes is the likeliest to contain something capable of displacement?

We can quickly dispose of the first two. Social signals wouldn't be social if they didn't involve manipulating the actions of other group members, something that can only be done in the here and now. Mating signals, apart from those that merely indicate species, sex, and/or availability, consist of advertisements of the fine genetic stock the advertiser comes from—displays of fine feathers, aerobatic skills, capacity to bear handicaps, ability to defeat rivals, or whatever. These qualities can only be exhibited in the present; there's no way any animal can convey any equivalent of "I may look a bit run-down at the moment, but you should have seen me last week."

This leaves us with survival signals, which again fall into two broad classes: alarm calls and food calls. Alarm calls we have already dealt with at length, and seen that they are inextricably linked to the appearance of predators or at least their presumed appearance (presumed by both caller and receivers if the caller mistakenly believes there's a predator there, and by the receivers even if the caller is lying to them). Food calls are mostly immediate reactions to the discovery of a food source, intended to be audible (or if signed, visible) to group members in the immediate vicinity. None of these is a likely candidate for displacement.

However, suppose that the food was some distance away from any other group member, and suppose that a measurable length of time had to elapse between the discovery of a particular food source and the transmission of news about that discovery to other animals. If any animal signal could be used in this kind of situation, wouldn't it qualify as an escape from the prison of the here and now, the first true case of displacement?

KINDS OF SIGNS

Surely it might. But what kind of signal would it have to be? So far, I've only mentioned two kinds: indexical and symbolic. But symbols can't just pop out of the woodwork; unknown in any known ACS, they would have to be worked for and won in the earliest stage of proto-language. And indices are irredeemably bound to the here and now, since they must point directly at whatever they refer to.

Fortunately, there is a third class, the iconic. An iconic sign is something that resembles what it refers to—in some way. It can be a part of the thing referred to, or a picture of it (or of part of it), or the noise it makes—anything that somehow evokes an object in the real world (or even an abstract class, as symbols do, it turns out).

I'm going to treat these three—icon, index, symbol—quite differently from the way Deacon treats them in *The Symbolic Species*. According to him, they form a hierarchy: icons at the bottom, indices in the middle, symbols at the top. Once or twice, when I've been thinking sloppily, I've endorsed this view. But viewed from the perspective of displacement, they're not a hierarchy at all.

First of all, let's make it quite clear that not all words are symbols.

Words can be iconic. There's nothing in "dog" or "cat" that, in and of itself, evokes those particular animals. But words like "buzz" and "hiss" evoke by mimicry the sounds they describe. Moreover, "buzz" doesn't necessarily refer to a particular noise on a particular occasion, in the way that iconic signs in ACSs do. A bristling of the fur is an iconic sign that can only mean a particular animal is angry at a particular place and time, but "buzz" may be applied indiscriminately, to refer to the noise made by the particular bee that's bugging you right now or the generic noise that bees, wasps, beetles, and other insects habitually make—even the similar noise a crowd of people in animated conversation makes.

Words can be indexical. "This" and "that" are used exclusively to point to particular objects in the world. Unfortunately, they're not very informative and can only be used if their referents have been established in the course of previous speech or writing, but they clearly can't stand in for a specific class of real-world things, like "dogs" or "tables," in the way a symbolic word can.

Then there are all the grammatical items that are also words: "the," "a," "who," "not," "by," "for," "of," and so on. These, unlike icons, indices, and symbols, do not even refer; they merely establish relationships between words that do refer. So a simple statement like "Words are symbols" just doesn't cut it.

It remains true, however, that most words are symbolic, and that without symbolic words we couldn't have language. So here it comes, the question that *The Symbolic Species* never answered (or even asked, come to that)—where did symbolic words come from?

Well, think back to "buzz" and "hiss." Compare these:

There's a mosquito buzzing in my ear.
Nothing is more irritating than a buzzing sound.

In the first, there's a particular buzzing in the here and now. In the second, there may be, but there needn't be—I could say this in reacting to a story about something that happened years ago. In talking about symbolism and words, people often make far too much of arbitrariness—the absence of any relationship between a word's form and its meaning. You almost feel there's some kind of class distinction in there—up top are the symbolic words, words whose meaning you couldn't possibly guess, and underneath the second-class citizens, words that shamelessly wear their meanings on their sleeves. But just as members of all social classes share the same bodily functions, so symbolic and iconic words share the same capacity for displacement. And when it comes to how language began, displacement is a factor far more important than arbitrariness.

Indeed, arbitrariness is a feature of many animal signals. To come back to the vervet's leopard call, there's nothing in it that, in and of itself, evokes or even suggests a leopard.

Rather than forming a hierarchy, symbols, icons, and indices can best be visualized as the points of a triangle. Indices, at one corner, cannot under any circumstances have a capacity for displacement. Symbols, at the remotest corner, must have that capacity. And icons, at a corner closer to symbols, may or may not have it depending on how they're used. Among other animals, icons never developed that capacity, but it was potentially there, and it blossomed when language came along.

Iconicity, therefore, is the most probable road that our ancestors took into language.

So we've now completed the task that, as language engineers, we set out to accomplish. We can specify precisely the conditions under which an ACS could begin to morph into something completely different, with different goals and different means for getting there. Here are the specifications:

Selective pressure: the need to transmit information about food sources outside the sensory range of message recipients.
Probable means: iconic signs.

Now the question becomes, can we find any evolutionary model for this? Are we talking about something wholly unique to some human ancestor, or is there anything comparable among other species that we could use to guide us here? If not, we run the risk, endemic to this field, of floundering in a morass of speculation, a morass with no empirical floor to prevent us from being swallowed by the mud. It's happened to all too many before us.

So let's stop thinking like engineers and start thinking like biologists. The comparative method is the core of evolutionary biology, and it's precisely the absence of anything language could be compared to that has bedeviled the field from the outset. It's time to start looking at evolution and how it has worked in the past, to try to make some linkage between our specifications and things that have really happened.

Where to start, though?

The obvious place, the place almost everyone starts, is among the great apes. After all, they're our nearest relatives, made from almost the same genetic material. If there's continuity in the development of language, that continuity surely should have a straightforward genetic basis of some kind.

Well, in the next couple of chapters we'll see about that.

3

SINGING APES?

THE IMPORTANCE OF BEING PRIMATES

By the close of the nineteenth century, everyone who wasn't blinkered by religious dogma knew that humans, far from being the special-purpose product of some interfering deity, were members of the primate family who could only have evolved from something resembling a chimpanzee. The full significance of this finding wasn't appreciated until in 1954 the discovery of the double helix structure of DNA ushered in the century of the gene.

As discovery followed discovery, genetics came to dominate the life sciences. The determinism of Richard Dawkins's *The Selfish Gene* went mainstream; while lip service was paid to the environment and its influence, a consensus grew that genetics formed by far the most important driving force in evolution. As the close resemblances between the DNA of humans and great apes became apparent, more and more people assumed that most if not all of what had been seen as typically human (and in many cases, uniquely human) traits and behaviors were no more than expansions of traits and behaviors found among the apes. Such a view was encapsulated in the titles of popular works on human evolution: *The Naked Ape, The Third Chimpanzee,* and (of course) *The Ape That Spoke* and *The Talking Ape.*

Here lie the roots of what Irene Pepperberg described as the "primate-centric bias" in language evolution studies. It seemed self-evident that if humans were, indeed, no more than souped-up apes, the origins of language, or at least its immediate precursors, had to be found among chimpanzees, gorillas, and orangutans. This belief was reinforced by attempts, from the 1960s on, to teach some form of language to these species (more on this in the next chapter). Although, to most linguists, results of these experiments were equivocal at best, the case made by "ape-language" researchers found many supporters in other behavioral sciences, and has been strengthened, over the past couple of decades, by the performance of another great ape, the bonobo or pygmy chimpanzee.

Certainly the simplest and most straightforward story of how language evolved would go something like this:

Five to seven million years ago, the primate line of descent (which had already branched to throw off first orangutans and then gorillas) divided again. One branch went on to father bonobos and chimpanzees; the other was that of our ancestors. These ancestors, driven out of the forest and into the grasslands by global drying, were forced to make meat a substantial part of their diet. A rich meat diet enlarged their brains and thereby increased their intelligence. Intelligence was also enhanced by a factor common to chimpanzees and bonobos—social competitiveness. Apes were always trying to outwit one another, to gain prestige in the group, preferential access to mates, first dibs in a monkey hunt. This factor selected for the intelligence that meat supplied, setting up a beneficent spiral. The old ape ACS, drawn into this spiral, expanded and diversified. At some point that it might be quite impossible for us to determine, the ACS flowed seamlessly into language. Language made life more complicated, and in turn itself grew more complicated to deal with these complications, until we finally arrived at our present situation.

To doubt that our language has its foundations in ape behavior has become, in some circles, almost a heresy, enough to brand the doubter a closet creationist. That this conventional wisdom has achieved the status of dogma, to be accepted on faith, regardless of the absence of evidence—even if what evidence there is points in a contrary direction—is made clear in a quote from Steven Mithen's recent book on the evolution of language: "And yet it is in this African ape repertoire, and not in those of monkeys or gibbons, that the roots of human language and music *must be found*. The *apparently* greater complexity of monkey calls *must be an illusion*, one that simply reflects our limited understanding of ape calls" (my italics). Note that, as well as demonstrating an abiding faith in the conventional story, this passage inescapably suggests a bias I mentioned in the last chapter: homocentricity.

I suspect most of those who subscribe to the conventional wisdom don't realize they're homocentric; most likely they'd indignantly deny it, pointing out that, instead of placing humans center stage, they're emphasizing all the things we have in common with apes. Maybe, but if you see apes as just slightly less complex versions of ourselves, you're implicitly saying they're merely a crude foreshadowing of humans—*In the Shadow of Man*, as Jane Goodall's book title so aptly put it. They're worthy of our attention only insofar as we can find in them forerunners of typically human traits—they're "almost human." And where language is concerned, it's as if you had something like the old *scala naturae* of pre-Darwinian nature study, a ladder leading up through progressively richer and more complex systems of communication until finally, among our nearest relatives, you found something that needed only a tweak or two to blossom into language, or at least protolanguage. It's an embarrassment to this whole approach that monkeys turn out to have things more wordlike than anything found among chimps and bonobos.

If ACSs formed a ladder to language, the species closest to us should have the most languagelike ACSs. But they don't, and the reason they don't is because ACSs are not failed attempts at language. They're not crude and misguided attempts to do what we do. They are autonomous systems that exist to serve the adaptive needs of each species that has one. Species get what they need and no more, and in the section after next we'll see good reasons why the ACSs of great apes should be, in some ways at least, less, rather than more, complex than those of monkeys.

SOME PROBLEMS WITH BEING PRIMATES

Let's look at some of the things that are wrong with the conventional story.

Many versions of this story have language arising through an intensification of social competitiveness—a kind of arms race between increasingly smarter and more devious protohumans. But this ignores two vital things. It ignores the ecology of our post-chimp, prelanguage ancestors, plus all the changes from typically ape social behavior that this very different ecology would inevitably impose. And it ignores the fact that those ancestors developed, at some stage, but likely a long time ago, a type and degree of cooperation unknown in any other primate species. It's not that humans aren't socially competitive—they are, and highly so. But, paradoxically, they are also highly cooperative, capable of combining their forces for joint ventures in units that range from dyads to many millions of individuals. Among apes, on the other hand, the dyad is about the largest unit in which cooperation can occur among unrelated individuals; even then, it's only of the "I'll scratch your back if you scratch mine" kind. Both ecology and the source of cooperation will be discussed in chapters 6 and 8.

Then there is the dubious nature of the connection between language and intelligence, an issue that will come up in several chapters, but particularly in chapters 4 and 10. At first sight, that brain link looks like a no-brainer. We're smart, we have language; other animals are less smart, and they don't have language. We assume that intelligence is the broader category. Since we usually regard language as no more than the means by which we express our thoughts, it seems natural to think that language should issue from intelligence, rather than vice versa. It seemed equally obvious, to naive observers, that the earth was the center of the universe, and the sun, moon, and planets all went around it.

When it comes to mind, intelligence, and language, we're just about where people were with regard to the universe, say a thousand years ago.

However, I'd like now to focus on the transition issue (that is, the transition from an alingual state to anything that you might, with a stretch, classify as the beginning of language) and examine some of the problems this transition must face if we adopt a purely primate-centric approach.

Most accounts gloss over this transition, one way or another. Where we need a precise, detailed, carefully reasoned analysis, we seldom get more than hopeful hand-waving. But the transition from no language to some kind of protolanguage is where the rubber meets the road—it's the crux, the core moment in language evolution, and the whole of the middle portion of this book will be devoted to what I believe, in our present state of knowledge, is the best and perhaps the only possible explanation of that moment. But first we need to see why a straight-line evolution from a typical ape ACS to language cannot be maintained.

We should begin by seeing just what we have to work with. Accordingly, we'll look at the current ACSs of bonobos and chimpanzees, assuming that these systems have not deteriorated or shrunk since their line split from that of humans. (We cannot, of course, know this for certain, but there's no reason to suppose otherwise.) Since, in the first two chapters, I've described what the most minimal language should be able to do, we can then assess how plausible it is that some expansion or complication of these ACSs might lead to anything you could call protolanguage.

THE RAW MATERIAL OF LANGUAGE?

One of the best recent accounts of chimpanzee and bonobo ACSs is by Amy Pollick and Frans de Waal of the Yerkes National Primate Research Center. They list a total of thirty-one gestures, fifteen vocalizations, and three facial expressions. Of these signals, three gestures and six vocalizations were produced only by chimpanzees, while two gestures and six vocalizations were produced only by bonobos. From these data Pollick and de Waal draw a rather surprising conclusion: support for the theory that human language was originally gestural rather than vocal.

This is surprising because if 84 percent of the gestures are shared by the two species, but only 20 percent of the vocalizations, the gestures are more likely to be homologies—traits directly inherited from the common ancestor of chimps and bonobos, rather than innovations appearing after the split. On the other hand, vocalizations, since they differ more in the two species, must have developed mostly since the two species split. And, since language too is vocal and an innovation, we would expect protohumans, just like apes, to have been moving away from their last common ancestor, not toward it—shifting from gesture toward a more vocal medium.

But the claim that language started in gesture is not the only dubious claim Pollick and de Waal make. They note that gestures are more contextually flexible than vocalizations—that is to say, the meanings of gestures vary according to the contexts in which they are used. For instance, the gesture described as "gentle touch" may be an invitation to sex if used by a male in the presence of a receptive female, but a request for milk if used by an infant to its mother. On the other hand, the "scream" vocalization is always and only given (by members of both species) if an individual is under threat or actual attack. They therefore argue that gestures, like words but unlike calls, are not tied to particular situations, and thus show movement away from the rigid restrictions of ACSs.

But at least the calls show consistency of meaning, whereas gestures have no consistent core of meaning—unlike words, they take their meanings entirely from the circumstances in which they are used. (Imagine, if you can, a word "moosh," which means "Let's have sex" if uttered by a male to a female and "Give me milk" if uttered by a baby to her mother.) Gestures differ from calls only in that they are linked to several different kinds of situations rather than just one, and mean something quite different in each situation. That's not how words work.

Moreover, the goals of all these gestures and vocalizations are anything but languagelike. They have two goals only. One goal is to express emotion. The "food peep" vocalization, for instance, is seldom if ever used to indicate the location of food to other group members—its "meaning" is "Yum yum!" rather than "Come and get it!" The other goal, as with the "gentle touch" discussed above, is to manipulate another member of the group. Monkey alarm calls at least give concrete information (or can be so interpreted), but not one call or gesture in the bonobo or chimpanzee repertoires transmits any objective information about the environment.

Thus, great ape ACSs are more limited in range than monkey ACSs. According to the ladder-to-language approach, it should be the other way around. But from the perspective of this book, what we find is exactly what we'd expect. Each species has the system it needs to do what it has to do. Monkeys need warning signals about predators because they are usually smaller than apes, frequently live in areas where they have to spend time on the ground, and are therefore more vulnerable than apes. The more specific the signal, the greater the fitness that results from it. Accordingly, the main classes of predator—aerial, terrestrial, serpentine—may be differentiated, giving rise to things sometimes mistaken for protowords and misinterpreted as "steps toward language."

Great apes, on the other hand, are large and strong enough to discourage most predators. They also live mostly in densely forested areas where at any time they can climb beyond any predator's reach (except for the gorilla, of course, which is too large, and can look too ferocious, to form anyone's prey). They don't develop alarm calls because they have so little to be alarmed about.

Considered alongside the uniqueness of language, facts like these lead to the conclusion that whatever caused language to blossom appeared not among any living species, but among some species ancestral to humans but long extinct that emerged after our line had separated from the chimp/bonobo line. Any such species would, of course, have had a lot in common with its ape ancestors. But not everything—by no means everything. Crucial differences in food sources, consequent scavenging techniques, and interactions with other species could have caused profound changes in behavior, including ways of communicating and things to be communicated about.

But here we must proceed with immense care. We know next to nothing about what immediately followed the split. All we have to work on are rare and scattered fragments of bone and a handful of extremely primitive tools. We must steer, like mythical navigators, between the Scylla of viewing these ancestors of ours as carbon copies of modern apes and the Charybdis of treating our ignorance as a license to print money—of drawing a picture of our ancestors based more on its intuitive appeal than on its plausibility.

Among all the theories of how language evolved, perhaps the most appealing of all is the singing ape.

Music maketh man?

Music and language are both universal in the human species and both are unique to it. Each is distinguished by having structure that is complex as well as rule-governed, and by being (in contrast with the songs of most other species) potentially infinite and open-ended. What would be nicer and neater than to find that they are intimately connected, and share the same origin?

In addition to its intuitive appeal, this notion has a long history. The idea that language and music are intimately connected in their birth goes back at least to Rousseau and other Enlightenment philosophers. To Darwin, it appeared "probable that the progenitors of man, either the males or females or both sexes, before acquiring the power of expressing mutual love in articulate language, endeavored to charm each other with musical notes and rhythm." According to Otto Jespersen, writing half a century later, "language was born in the courting days of mankind—the first utterances of speech I fancy to myself like something between the nightly love-lyrics of puss on the tiles and the melodious love-song of the nightingale." A more thorough, but in parts equally poetic, development of the notion has appeared most recently in *The Singing Neanderthals*, by Steven Mithen of the University of Reading. Note that the connection between sex and what has been called "musilanguage," found in the work of Darwin and Jespersen, fits well with the idea, favored by Mithen and Geoffrey Miller, that language may have originated, at least in part, as a form of sexual display.

One problem with this view is that, when it comes to our nearest relatives, their vocalizations are as lacking in music as they are in objective reference. To find anything that looks like a musilanguage precursor, you have to go back as far as the gibbon.

Gibbons are middling close relatives of humans; they split from the common ancestor of the great apes and us somewhere between twelve and twenty million years ago. And gibbons sing, no question of that. Their songs can last up to an hour and a half, longer than most human songs. Moreover, gibbons engage in duets, almost always performed by mated pairs (gibbons, in contrast to great apes, are highly monogamous). So it has been suggested that some unspecified human ancestor developed similar singing routines. At some subsequent date (in Mithen's version, as late as the emergence of modern humans, a mere couple of hundred thousand years ago) musilanguage, or what Mithen also calls "Hmmmmm"—holistic, manipulative, multi-modal, musical, and mimetic utterances—would have split into something that became music and something that became language.

In general (except for the aquatic ape hypothesis, which claims that at some stage of their evolution human ancestors lived at least partly in water) evolutionary science prefers blanket explanations—they have the irresistible appeal of two-for-one offers in your local supermarket. If a single umbrella theory will explain a variety of traits, that's almost always preferred to a series of different explanations for each trait. But here is one case in which, for a variety of reasons, an umbrella theory won't work.

Such a theory would require that loud and prolonged singing of some kind was performed throughout the period, well over a million years in length, when human ancestors subsisted in a largely treeless savanna, drier and more extensive than the savannas that are presently found in parts of East Africa. Why would prehumans have practiced this behavior, under these conditions?

Well, why do gibbons sing?

Gibbon authorities suggest several main functions. One is pair-bonding; the long and repeated exchanges of the duets are supposed to enhance solidarity between mates, and they seem to work, since gibbons are among the mere 3 percent of mammals that practice monogamy. Another is territorial boundary marking, the warning off of potential trespassers—mated pairs of gibbons divide the rain forests into sharply defined and firmly defended territories. A third is simply keeping in touch with one another, in places where the denseness of the canopy renders one invisible at ranges of more than a few yards.

But what could possibly have been the functions of song for a prehuman species in largely treeless grasslands?

Song as a pair-bonding mechanism is highly unlikely. Human ancestors probably weren't monogamous—great apes aren't, and neither are we, even if we try or pretend to be, so a monogamous interval at any time in the past looks unlikely. But suppose we did go through a monogamous period. If two mates don't happen to be out of sight of each other up two different trees, there are countless more effective ways of bonding than yodeling at each other.

Human ancestors probably weren't territorial, either—at least not in the sense of holding small, well-defined chunks of territory. Most likely they had a fission-fusion social structure, like that of contemporary apes; that's to say groups would be continually splitting up and reforming, merging with other groups. In open terrain, where different groups might utilize the same areas at different times without conflict or even contact, what would be the point of noisily defended frontiers?

Furthermore, the terrains in which gibbons and human ancestors lived were such that for maintaining contact sound was essential in one but useless, even dangerous, in the other. Until relatively recently, humans were not forest dwellers, and our ancestors lost the capacity to live in trees when they became fully bipedal, millions of years ago. Gibbons, on the other hand, spend their whole lives in trees, brachiating around with a speed and dexterity unique and quite remarkable for primates of their size.

The length and loudness of gibbon songs result directly from this environment. Their forests are dense, but their populations aren't; they move about in their pairs, and one pair seldom meets another. Even members of a pair often can't see each other, so they would lose contact with mates and blunder into other gibbons' territories if they didn't make a lot of noise a lot of the time. Moreover, they can make noise with impunity—predators below may hear them, but are hardly likely to go after them into the canopy.

But on the savanna, where there were beasts with keen hearing far larger and more lethal than our ancestors, to sing out with any frequency would have been to write one's own death warrant. Moreover, the absence of trees and the level or undulating nature of most savannas mean that, in contrast with the rain forest, animals are visible at considerable distances. To be out of sight is, under those conditions, almost always to be out of earshot—there's little point in yelling and hoping your friends will hear you.

To assume that, even if our ancestors had sung before, they would go on singing under these conditions is absurd—something you can do only if you think that behavior and environment are completely divorced from each other. But behavior and environment aren't watertight compartments—they're intimately linked; they shape each other into a lock-and-key fit. And this fit is precisely what is meant by "adaptation"; we'll see exactly how this works when we come to chapter 5.

Conditions on the savanna were such that while they lived there, our ancestors very probably produced *less* sound than our ape relatives, not more. If this was indeed the case, a single source for music and language becomes highly unlikely. Unless, of course, someone succeeds in coming up with some function prehumans had to perform, under those same savanna conditions, that they couldn't have performed by any means other than by singing. It's unlikely anyone will, but never say never in science.

Meanwhile, the sinking theory of musilanguage has recently received a life preserver flung from a quite different direction, one that seems to resolve a completely different problem the theory faces.

This problem involves the mechanics, the nuts and bolts of how you might be able to transition from song to language. It's the problem of how something that has no meaning acquires meaning—not the vague, general emotional sensations that music arouses, but precise, specific references to things. The proposed solution wasn't designed to support the singing-ape hypothesis, but it has been eagerly grabbed by musilanguage advocates like Mithen because, just like a life preserver, it does seem to keep them afloat in a very choppy sea. Let's see how and why this solution was proposed.

The appeal of holistics

Note first that the problem of how to get from songs without words to some kind of protolanguage is an artificial problem. By an artificial problem I mean a problem you don't need to have, that you create for yourself—one you could avoid by simply abandoning the singing-ape hypothesis. And it's a problem you can only solve by proposing a kind of protolanguage quite different from what most people had previously envisioned.

When I first developed the idea of protolanguage, I saw it as being something like a modern pidgin, a pidgin in its earliest stages. It would have consisted of a small number of wordlike things—whether signed or spoken is of no importance; there would quite likely have been both kinds—strung together haphazardly, if at all, without anything you could call grammatical structure, and eked out by pointing, pantomime, and any other device that might come to hand or to mouth.

These wordlike things wouldn't have been like today's words in their form, of course. For one thing, every one of today's words, in whatever language, is composed of from one to a dozen or so highly specialized but in themselves meaningless speech sounds, each drawn from an inventory of from eleven to a hundred-odd sounds, depending on which language you're speaking. In contrast to this, the words of protolanguage, even if vocal, could not have been divided into component parts, and would likely sound to us like meaningless grunts or squawks. But, like today's words, each would have a fairly well-defined range of meaning, and that meaning, rather than relating directly to the current situation, would refer to some relatively stable class of objects or events, regardless of whether or not these were present at the scene.

By adopting this notion of protolanguage, you reduce the basic problems of explaining how language evolved to just two—explaining how words evolved and explaining how grammatical structure, syntax, evolved. You don't have the additional problem of explaining a further intermediate stage between nonlanguage and language—in this case, singing. And the singing-ape hypothesis did nothing toward resolving the other problems, the origins of words and the origins of syntax. Even if you could explain how singing evolved, you still had to explain these two.

The idea of the kind of protolanguage I proposed was widely accepted through the 1990s, even though it didn't help to explain where words came from in the first place. And then Alison Wray, of the University of Cardiff in Wales, came along with another highly appealing idea.

Wray did not subscribe to the singing-ape hypothesis, but what she proposed looked like it might solve the transition problem for singing-ape supporters. It also seemed to offer assistance to computational modelers of evolution and to strong continuists who believed in a straight line of development from ape ACS to language, so it had plenty of traction from the get-go. Here is what she proposed.

As we've seen, ACS units are holistic. They correspond in meaning not to words, but to whole phrases or sentences: "Stay off my territory!," "Mate with me!," or "Look out, predator coming!" They are holistic in the sense that there are no parts of whatever corresponds to "Mate with me!" that mean, individually, either "mate" or "me." Moreover, the number of these units is more or less fixed, for any given species, and they are not learned, but (in some sense) innately programmed.

Suppose that, in some species ancestral to humans, the latter two restrictions were lifted. Suppose that this species could add indefinitely to its ACS repertoire—could invent and learn a whole series of holistic units that would similarly serve to manipulate the behavior of other group members (remember the consensus born in the late 1990s that a growing sophistication of social intelligence was the main driving force behind language). Eventually, the inventory of holistic signals would get so big that it would impose too heavy a load on memory. Fortunately, before the strain became intolerable, another development would have taken place.

Suppose that the holistic signals were phonetically complex—in other words that they consisted of a number of segments that could be distinguished from one another. Two of Wray's examples are hypothetical holistic calls, "tebima," meaning "Give that to her," and "matapi," meaning "Share this with her." (Why anyone would develop two holistic calls quite different in structure that would largely overlap in meaning is one of the things about a holistic protolanguage that remain unexplained.) These calls happen by pure chance to share a syllable, *ma*.

The sharing is, of course, coincidental; there is nothing about *ma*, or any of the other segments of these calls, that yields any kind of meaning in itself. In Wray's own words, "The whole thing means the whole thing." However, the double coincidence—that the syllable occurred in both calls, and that both calls contained reference to an unspecified female recipient, or potential recipient—would be picked up by some smart hominids. They would then begin to use *ma* as a signal for "female recipient," joining it with other fragments similarly gleaned, to start building a stock of words. And that is how language took on the compositional structure—isolated words having to be put together to form sentences—that we know and use today. Instead of starting with words and building them into sentences, you started with sentences (or rather the semantic equivalent of sentences) and broke them down into words.

This is a highly ingenious proposal, particularly in the way it seems to offer a viable bridge between ACSs and language. Unfortunately it raises a host of objections, only a few of which—hopefully the most serious—we'll have space for here.

WHAT'S WRONG WITH A HOLISTIC PROTOLANGUAGE

Let's grant the wholly unsupported assumption that these holistic calls would have been structured in such a way that you could actually divide them into segments—that you could say, "Here's a subunit within the unit that starts here and stops there." In fact I doubt there's any existing animal call that you could do this with, or if you could, that you'd find anyone to agree with you on where the boundaries lay. But, for the sake of argument, let's suppose this was possible.

First of all, there'd be the problem that while *ma* might occur in two holistic calls both involving female recipients, it would also occur in other calls that had nothing to do with female recipients. Fine, say the holistic defenders; the other cases don't matter, once you've spotted a sound/meaning coincidence, you stick to that and just ignore all the other calls in which *ma* occurs—*ma* now means "female recipient" and that's that. Well, let's be generous and grant that too.

The real problem is, as I said to Steve Mithen at a recent meeting, that in order to extract segments from a holistic signal, you'd first have to know English. Steve got very upset at this; he thought I was being facetious. Provocative, maybe, that's my style, but facetious—no way.

You see, the whole holistic proposal depends on the assumption that for every holistic signal, there's just one, and only one, equivalence between a holistic signal and some sentence in English (or another human language). If there isn't, how can you possibly agree on what all the bits of the signal really refer to? Take the vervet alarm call for eagles that we looked at in chapter 2. As we saw, this could equally well be translated in at least three different ways: "Look out, an eagle is coming!"; "Danger from the air!"; "Quick, find the nearest bush and hide in it!" Suppose that the call could indeed be broken down into two or more segments. Unless our holistic hominids somehow already knew the equivalent for the call in English (or some other human language), how could they assign an unambiguous reference to those segments? Would the segments be taken to mean "eagle" and "coming," or "danger" and "air," or "bush" and "hide"?

We cannot simply assume that what we might choose to describe as "the eagle call" or even as "something like a word for eagle" necessarily refers to eagles. ACS units aren't designed to refer; they're designed to get other animals to do things. Such units don't really translate into human language. We can give an approximate meaning, or several possible meanings, in terms of our language, but the idea that underlying both is the exact same semantic expression is simply baseless. And this absence of common ground is not something that happens to be so; it's something that has to be so.

The deepest reason why a holistic protolanguage is impossible is the profound fallacy that underlies Alison Wray's proposal. Once again, it's the fallacy that language and animal communication are fundamentally the same kind of thing. Animals were struggling toward language, but the poor schmucks just weren't smart enough to make it—we, on the other hand, were. This belief, tacitly and probably quite unconsciously held by many who see themselves as fighting anthropomorphic biases, is in fact hopelessly anthropomorphic itself. It should therefore be quickly abandoned by them once they realize its true nature.

A holistic protolanguage, even if it could exist, wouldn't really be a protolanguage at all. It would be in some ways like a hyperinflated ACS, a series of reactions to specific situations; it would, however, be anomalous as an ACS, insofar as the situations it reacted to would seldom be fitness-enhancing ones. It would in fact be a hybrid, neither one thing nor the other, yet still not a viable intermediate stage between ACSs and language.

A holistic protolanguage assumes that the parts that make up an ACS are like the parts that make up language, only finite instead of infinite and instinctive rather than learned. Make it so those parts become open-ended and learnable and you've taken a giant step toward language—so goes the holistic claim. But that cannot be, because ACS units and language units are totally different in form, in function, and in anything else you can think of.

Moreover, unlike the kind of compositional protolanguage I had proposed, a holistic protolanguage would not be able to do even one of the most basic things language can do. You could not use it to ask questions or make negative statements. You could not hold a conversation in it. You could not use it to provide any kind of novel information. All you could do with it is what you do with an ACS—manipulate people.

Wray's objection to my kind of protolanguage is that it would have been initially crude, highly ambiguous, and not much use in manipulating people. Fine, right on the money—manipulating people wasn't its job. For that, we already had, and we still have, a perfectly good ACS. We have screams, tears, laughter, rage faces and play faces, presentations of finger and/or rump, a plethora of body language to show how we feel and what we want others to do about it.

An ACS is one thing, language another. If there's any way to get from an ACS to language, it can't consist of just blowing up the ACS until it almost bursts, then hoping the pressure will make it morph into something completely different. If the transition was ever to be accomplished, it could only come through introducing into an ACS some alien factor, something like the piece of grit that, when inserted into a humble oyster, produces a magnificent pearl.

BABIES TO THE RESCUE?

The singing-ape hypothesis is not the only appealing idea that tries to solve the language-origins problem by turning the meaningless into the meaningful. A variant on it has recently been proposed by Dean Falk of Florida State University, one that involves mothers and babies. Again it's an ingenious and seemingly plausible idea, but again it founders when we start asking how meaning crept in.

The best thing about Falk's proposal is that, unlike the singing ape, it's based on a real evolutionary development unique to the human line. (Remember that starting from such a development was listed in chapter 1 as one of the minimal conditions any valid theory of language evolution has to meet.) This development occurred when brains began to enlarge.

Normally, mammals are born with their brains fully developed—everyone who's seen a calf or a lamb born must have marveled at how soon after birth they're able to do things like walking that take human infants a year or more to acquire. But a fully developed human brain just wouldn't go through the birth canal. Evolution therefore favored mothers who brought their babies to term when brain enlargement was far from complete. The downside to this: babies are helpless for several months after birth, and need a mother's fairly constant attention for several months more. Which presents a problem if Mommy simultaneously has to forage for her (and the baby's) living. If you doubt this, try picking berries (or worse, digging up roots) while you're holding a squirming baby.

So how would she control baby? She'd have to put it down. Falk pours scorn on the notion that baby slings were an early invention; she says if you believe that, go into any woodland and try making one out of natural materials (and that's without even taking into account that in the savannas there weren't any woods to go into). There might later, when hunting of larger animals began, have been slings made out of animal hides, but the problem of the terrestrial toddler arose long before that. On the ground, back in the Pliocene, there was limitless trouble a crawler or a toddler could get into. The only thing that would work, according to Falk, would be some kind of vocal communication.

Now no one could dispute, in this context, the value of sounds of reassurance, hushing sounds to keep baby quiet when there were predators around, shouts of warning when the little fellow was about to put a poisonous berry into its mouth, and so on. But why did these sounds have to develop into meaningful words? Wouldn't meaningless sounds that were pleasing (for reassurance) or alarming (for warnings) and simply stayed that way have done the job equally well?

Falk has never addressed this issue head-on. Instead, when she published her theory in the journal *Behavioral and Brain Sciences*, she sidestepped the problem by saying, "As is true for human babies towards the end of the first year, prosodic (and gestural) markings by mothers would have helped early hominin infants *to identify the meanings of certain utterances* within their vocal streams . . . Over time, *words would have emerged* in hominids from the prelinguistic melody and become conventionalized" (my italics).

Behavioral and Brain Sciences is rare among scientific journals in that every article appearing in it is followed by commentaries from a couple of dozen or so scholars from the various fields that the article involves, and includes a response to those commentaries by the article's author. In my commentary on Falk's article, I pointed to the passage I just quoted, and remarked that it provided virtually all Falk had to say about the transition from meaningless infant-controlling noises to a meaningful word-based protolanguage. But how could infants "identify the meanings of certain utterances" unless the meanings of those utterances were known to the mothers who uttered them? And how could the mothers have learned those meanings? Well, from *their* mothers. And how would their mothers have learned them?

See where I'm going? Right—infinite regress. If we take Falk literally, language has to have been there since the beginning of time. If we don't take her literally, what could she mean? That "words would have emerged"? How? Why? They just popped out one day?

The kind of utterances Falk is talking about have all the properties of ACS calls and none of the properties of words. They directly affect evolutionary fitness. They're not symbolic. They show no signs of displacement. They're meaningless outside of the contexts in which they're uttered. There's no way they could have brought us a tad nearer language.

In her response to commentary, Falk responded to a couple of other points I made, but not to the one about the transition to language. There's no way "words would have emerged" without some particular, highly specific set of circumstances that forced words to emerge—not from the singing of apes, not from the cooing of mothers, not from grooming, not from the decomposition of holistic utterances, not from any of the dozens more proposals that have been put forward over the years.

Wait, you say. What about the attempts that have been made, over the past four or five decades, to teach language to apes? Why waste time wondering how language began if in fact apes already have language—or at least are able to acquire it?

Their so-called language abilities deserve a chapter to themselves.

4

CHATTING APES?

On Saturday, August 24, 1661, Samuel Pepys, diarist, top-level bureau-crat, and inveterate womanizer, was taken to see "the strange creature that Captain Holmes hath brought with him from Guiny." Pepys was told it was a "great baboon," though it was more probably a chimpanzee or gorilla, species at that date still unidentified by Europeans. And Pepys, with no background whatsoever in ethology, ecology, primatol-ogy, psychology, or linguistics, had an insight that it would take science three centuries to catch up with: "I do believe that it already under-stands much English, and I am of the mind it might be taught to speak *or make signs*" (my italics).

About speaking he was wrong, of course. In the late 1940s, Keith and Cathy Hayes of the Yerkes National Primate Research Center, after years of trying, got their chimpanzee Viki to produce just four words: "mama," "papa," "cup," and "up." (I should make it clear from the very beginning that when I use the term "words" in connection with animal learning, I'm not in any way claiming these were words in the full human sense—"protowords" would be better, but since that's awkward and ugly I'll go on calling them words and ask you to remember this caveat.) An ape's vocal control and physiology just aren't capable of speech, although its hands, as Pepys knew (he was perfectly familiar with the sign language of the deaf) were well able to cope with the mechanical side of language. Another decade had to go by before a sec-ond couple, Allen and Beatrice Gardner of the University of Nevada, tried what should have been, but clearly wasn't, the obvious.

I certainly don't want in any way to detract from the Gardners' groundbreaking achievement, but there's a point I've never seen made that needs to be made. Their experiment was low-tech—none the worse for that, but the point is, it could have been carried out at almost any time in the past, and certainly by the second half of the seventeenth century. After all, that was the heyday, the first flush, of science and the empirical method. The year before Pepys's insight, the Royal Society of London ("for the improvement of natural knowledge") had been founded. While Pepys was watching his ape, Robert Boyle's book on how the human body uses oxygen was on sale in neighborhood bookstores. Two years later the first reflecting telescope was invented. A few years after that, van Leeuwenhoek discovered bacteria and the first measurement of light speed was made.

A century rolled by. Now we're in the Enlightenment, so-called, the era in which anything and everything could be questioned—including, naturally, the origin of language. And questioned it was: in 1769, the Royal Prussian Academy of Sciences offered a prize for the best essay on it. In the essays poured, all thirty-four of them, all right off the tops of their composers' heads. It did not occur to any of those scholars to see what, if anything, other species could learn. Or to any of the French philosophes, Condillac, Diderot, Maupertuis, Rousseau, who also dabbled in the topic—except one, de La Mettrie, author of the notorious *L'Homme Machine*, who thought that a trained teacher of the deaf might be able to teach sign language to an ape. But he didn't actually get one to try it, nor did anyone else.

Then came the Romantics; "wolf-children" were "in," and Jean Itard spent years trying to teach language to the Wild Boy of Aveyron. It's ironic to think what he might have accomplished if he'd devoted the same kind of energy to teaching sign language to an ape. And mind-boggling to think how different the behavioral sciences might have become, if someone had taken Pepys's hint three and a half centuries ago.

So why wasn't it done?

The answer can probably be found in the words of Max Müller, professor of classical philology at Oxford, who responded to Darwin's claim of the primate ancestry of humans by pontificating that "the one great barrier between the brute and man is *Language* [his italics]. Man speaks, and no brute has ever uttered a word. Language is the Rubicon, and no brute will dare to cross it." Müller was, of course, a confirmed

Christian, a Lutheran outraged at the suggestion that, by teaching about Hindu religion, he had somehow undermined Christianity. His words reflect an all-but-universal mind-set that saw humans as distinct from the rest of creation—one that effectively prevented anyone (save a few like the atheist de La Mettrie, who thought there was no such thing as the soul) from even thinking that Müller's Rubicon could be crossed.

So it wasn't until the scientific mind had been saturated by a century of Darwinism that it became possible for the Gardners to think the unthinkable, and for one "brute," the chimpanzee Washoe, to dare Müller's Crossing.

It was both a strength and a weakness of the "ape-language" experiments that so many who took part in them were driven by an agenda. Absent that agenda of not merely crossing but erasing the Rubicon—showing that no line could truly be drawn between apes and humans—the attempt might never have gotten off the ground. At the same time, the fact that the agenda was as much an article of faith as the religious conviction it aimed to overthrow drove advocates to rash and excessive claims. And this in turn led to an escalating series of controversies that, for many years, prevented any calm and objective assessment of exactly what those experiments revealed.

"ALMOST HUMAN" OR ONE-TRICK PONY?

On both sides, claims were carried to extremes. To some critics, signing apes (or their successors who were taught with some form of lexigrams, arbitrary and holistic representations of words that the animal could point to or touch) were, wittingly or unwittingly, conspirators in some kind of fraud. They were compared to Clever Hans, a horse in early-twentieth-century Germany who could answer mathematical and other questions posed by his trainer. So long as the answer could be given in terms of a number, Hans would provide it; he would stamp his hoof until he reached the correct number, then stop.

Oskar Pfungst, the psychologist who investigated Hans, found that the miraculous horse was picking up subliminal cues from his trainer's body language. As Hans approached the critical number of stamps, the trainer quite unconsciously grew tenser, then abruptly relaxed the moment that number was reached. Replace the trainer, or have him ask a question whose answer he didn't know, and Hans would stumble, his performance falling below chance. The totally involuntary nature of these clues is shown by the fact that on occasion Pfungst himself caught himself inadvertently giving away the answer.

A few of the earliest experiments may have been vulnerable to Clever Hans charges, but the Gardners and their successors soon developed techniques that ruled out unconscious cueing. What they did not, unfortunately, rule out were excessive claims by the experimenters that brought negative reactions not just from jealously turf-guarding linguists but from many impartial observers who would have accepted more modest assessments. Here are some examples of this Rubicon-fording hubris:

> "Washoe learned a natural human language."
> "Apes appear to be very similar to 2 to 3 year old human children learning to speak."
> "Koko has learned to use American Sign Language—the very same sign language used by the deaf."

These claims are simply untrue. First of all, "natural human language" contains at least two major ingredients that even the most sophisticated ape has never employed, let alone mastered. One is grammatical structure, or syntax, the complex set of rules and principles that determines whether a string of words constitutes an acceptable sentence in a given language or is simply a string of words (the odds that it will be the second of these, for any randomly chosen string, are tens of thousands or even millions to one, by the way). The other is grammatical items, all the "at"s and "did"s and "for"s and -eds and -ings and -ses that serve as signals of that grammatical structure, enabling us to parse and understand it rapidly and automatically, without a smidgen of conscious thought going into what is, let's face it, a frighteningly complex process.

These are not extra doodads indulged in by spoken languages but dispensed with by the deaf. Sign languages have just as rigorous a structure as spoken languages and just as many grammatical items to signpost that structure. Many of these grammatical items involve subtle body movements and facial expressions that the hearing observer, fascinated by flying fingers, simply misses.

Indeed, the claim that apes learned American Sign Language is absurd. They were taught a handful of signs equivalent to common referential words, and that was it. Some observers have described it, pretty accurately, as "pidgin sign." The apes never acquired the structure of ASL. Researchers would have been better off admitting that; then they would have been able to see something the apes *did* do that was, as you'll see, significant and surprising.

So even to claim that apes were "very similar" to two-to-three-year-old children is misleading. To begin with, very shortly after age two the average child transitions from short, unstructured word strings to complete, albeit still short, sentences. Soon after that, many children still several months short of their third birthday begin to produce several different types of complex sentence, using a range of prepositions, auxiliary verbs, determiners, and other grammatical items that no ape has ever produced.

Ah, say those who believe apes have, or can acquire, language. You're talking about production. We're talking about comprehension. Apes can comprehend language at the level of at least a two-year-old, and we can prove it. And in fact, they claim, comprehension is much harder than production. After all, in comprehension you have to figure out what the other person means. Whereas in production, you know what you mean, you just have to put it into words.

To a linguist, this is a bizarre position, the very reverse of the truth. The idea that we decide to say something and then dress it in words is one of those ideas, like the sun going around the earth, that seem obvious and irrefutable to the naive, untrained mind, but bear no relation to what actually happens in the real world. Even if the naive view were right, you'd still need the nuts-and-bolts, highly specific and detailed knowledge of how to put sentences together, in whatever language you were trying to speak, and how to do this smoothly and swiftly so your audience didn't get bored and wander away before you'd finished. In comprehension, on the other hand, you don't need to know how to put sentences together. If you know what enough of the words mean, and you know where you are and what's happening, and you can apply your common sense and practical knowledge of the world, you don't need syntax to figure out what the other person means.

If for instance someone says, "Go to the refrigerator and get an orange," you don't have to know that this consists of two coordinate clauses, that "to" introduces a locative phrase, or that "orange" is the direct object of "get." These are all things you'd have to know, in some sense, at some quite unconscious level—not the names for them, but what those names represent—if you were going to produce the sentence. To understand it, all you need are the meanings of four words: "go," "refrigerator," "get," and "orange." "Go" tells you about moving to another location—"refrigerator"—and "get" means you have to obtain something—"an orange."

So when Sue Savage-Rumbaugh pitted her bonobo, Kanzi, against a female toddler, Alia, on a series of commands like the refrigerator sentence (but varying the structure and content, naturally), Kanzi scored correctly 72 percent of the time against Alia's 66 percent. But did this really show that they were at a similar level of development? Kanzi was age eight and had had years of experience in hearing and executing instructions like these. Whether Alia had any experience at all is unclear from published accounts, but since testing started when she was barely eighteen months old, she could hardly have had more than a few weeks, at most. Add to that the fact, reluctantly admitted by the experimenters, that Alia's MLU (mean length of utterance, measured by the number of meaningful units—words and affixes—in a sentence) went from 1.91 to 3.19 over the six months of testing, while Kanzi's remained stubbornly stuck on 1.5, and you can see this experiment, impressive though it might seem, hardly shows parity between ape and child.

But statistics and formal measures don't really get to the heart of the matter. The real difference lies in content—not how apes and children communicate, but what they communicate *about*.

Ape conversation is ego-centered. All that any ape, including Kanzi, the Einstein of apes, ever talks about are things like where they want to go, what they want (or want you) to do, or what they'd like to eat. General topics are out. Objective information about the environment or events in it is never exchanged. And this, after all, is exactly what you'd expect from an animal without a natural language but with a fully functioning ACS. The things apes talk about, their own wants, needs, and desires, and the manipulative way in which these are expressed are, as we saw in earlier chapters, just the things that an ACS deals with, that an ACS is specially designed to deal with.

Contrast with this the behavior of Seth, the child whose brief flirtation with serial verbs I described in *Bastard Tongues*. When Seth was around the same age as Alia at the start of her tests—eighteen months— his father, one of my students, made a recording of a conversation he was having with a couple of friends, which Seth, still in the one-word stage, insisted on interrupting, even though no one was taking the slightest notice of him. Unfortunately I no longer have the transcript, but my memory of it is vivid enough to give you the flavor of a brief extract:

ADULT: Blah blah blah blah blah.
SETH: Telephone.
ADULT: Blah blah blah. Blah blah blah blah.
SETH: Fan.
ADULT: Blah blah blah blah blah blah blah.
SETH: Doggie.

Seth was systematically naming all the objects in the room. Obviously this was manipulative and self-serving—he wanted to join in the grown-ups' conversation. What's striking is the way he chose to do this: by showing the grown-ups all the things he knew and could recognize and had names for.

And that, as we saw in the last chapter, captures one of the most essential differences between the uses of language and the uses of an ACS. In using an ACS, manipulation is uppermost and information, if any, is incidental; in using language, information is inescapable—the mere fact of using language automatically transfers factual information from one individual to another. Seth surely was trying to break into the conversation. How he stated that need, though, was not through some ego-centered demand, as an ape would have done, but by informing everyone of the names of things—thereby showing that he now knew the human code, and was therefore entitled to join in the conversation.

So what can apes really do?

If the ape-language researchers hadn't been seduced by the excitement of dramatic claims and high-profile media impact, they might have been quicker to realize the full significance of what the apes were actually doing.

Most or all of the apes seem to have done at least three things, hardly mentioned in the literature, that were highly significant for any understanding of how language could have begun. I'll deal with these in ascending order of importance.

The first significant thing apes did was to distinguish between words and proper names.

To us, the distinction is self-evident. If I introduce you to a round-faced, short-bearded, spectacle-wearing individual named Rudolph, you don't start calling every round-faced, short-bearded, spectacle-wearing individual Rudolph. Similarly, if I show you a new kind of fruit and tell you it's a cherimoya, you don't, on seeing another such fruit in the market, ask the stallholder, "What do you call that one?" But why should the distinction between things like "Rudolph" and things like "cherimoya" be self-evident to an alingual ape?

Surprisingly enough, it apparently was. I say "apparently," because I don't remember seeing anywhere in the ape-language literature any specific discussion of this point. However, in a fairly extensive reading of that literature, I've seen no mention of any case where an ape called one of its trainers by another trainer's name, or where, on having been taught a sign for, say, "banana," the ape failed to apply it, or showed puzzlement when a banana different from the original training banana appeared. In other words, they seem intuitively to have grasped the difference between words for individuals and words for categories.

I don't know why this is so—again, no one studied it—but I would surmise that it comes from being a social species. You meet, and have to deal with, other members of your social group on an individual basis. You behave toward each of them in a different way. But you behave to every banana in the same way. In other words, the distinction is one that all social animals, simply by being social, get for free.

The second significant thing they did was to spontaneously put signs together.

As we'll see in a moment, it took them a long time to grasp what signs were about, even though they were being trained intensively. But they began to put signs together, to make messages, without any explicit training at all. They got some modeling, because their trainers would address them with sequences of signs, but not a drop of explicit instruction.

Granted, the combinations they made were seldom of more than two signs, which accounts for Kanzi's 1.5 MLU. But, lacking anything in the way of syntax, what would you expect? Granted, some of these messages were simply "X and Y"—two quite disconnected signs that happened to be given in sequence. But enough of them took the form "X[Y]"—true predications, things like "Roger tickle," "trailer go," "no balloon"—for these not to have been mere accidents.

Combining things, you'll remember, is something that animals using an ACS simply cannot do. There's no precedent for it in the animal world. How did they do it? I'll postpone that until we come to where our ancestors faced the same problem.

The third and most important of the significant things apes did is something I can only describe as "getting it."

The idea that an arbitrary symbol—be it a spoken word, a manual sign, or, as was increasingly the case, a pictorial symbol on a screen that the ape had to touch—can stand in for something in the real world is as plain as a pikestaff to you and me, but not to any member of another species. The apes took a long time to grasp it. Washoe took three months to acquire her first sign. Lana, an ape trained by Duane Rumbaugh, then at the Yerkes National Primate Center, took 1,600 trials to learn symbols for "banana slices" and "M&Ms." After that, however, they took off. Washoe was soon down to ten trials or less. Lana "succeeded in naming *ball* on its first presentation." It was as if a lightbulb had suddenly gone on in their heads: "So *that's* what these dumb humans are trying to get me to do!"

Well, a lightbulb going on in their heads is probably pretty close to what actually happened.

When we're talking about behavior, we tend to focus on what's going on in the external world, rather than on what must be happening inside the head of the behaver. If we think of anything in there, we usually speak of the behaver's "mind," and we probably think of that as some kind of cerebral screen, on which, as on the wall of Plato's cave, outside events are shadowed. Due perhaps to a hangover from centuries of dualism, we tend to ignore or downplay purely physical events in the head. But through the rest of this book, we must increasingly bear in mind the never-ending interaction between the external-physical and the internal-physical. What happens in the outside world triggers electrochemical events in the brain—sends messages racing down axons, enzymes leaping across synapses—but it doesn't only do that. It changes the way the brain is wired. And the long delay between the initial presentation of signs to apes and their first grasp of meaning follows directly from the following axiom:

"Neurons that fire together, wire together."

This is Hebb's Rule, a pithier version, known to all first-year neurology students, of a more nuanced but canonical statement by Donald Hebb, one of the pioneers of cognitive science. The brain's plasticity is hard to exaggerate. It can't do major architectural changes, but it can remodel most of its many rooms while it keeps the house running without the slightest glitch. Exactly what happened in Washoe's and Lana's brains, what hitherto weak or nonexistent links were forged or strengthened, we simply don't know. Nobody ever asked. Only years of neurological research could answer that question, and the ape researchers weren't even neurologists.

But we know in essence what must have happened.

The presentation of novel signs to apes, coupled with the presentation of physical objects, caused certain neurons to fire simultaneously that had never fired simultaneously before (this is true of any new experience). These neurons were those in the visual cortex that directly responded to the first sign and those (probably in either the motor or visual cortex) that represented the sign or written symbol that the apes were learning. (An interesting line of research that I don't think has yet been followed would be to confirm what one might expect: that apes stored representations of manual signs in the motor cortex but those of pictorial symbols in the visual cortex.)

It took repeated exposures to the signs, hundreds and thousands of them, for those same neurons to keep on firing together, each firing spreading and strengthening new connections until they were strong and far-reaching enough for enlightenment to dawn. That accounts for the long initial delay between first presentation and "learning." But once the first few signs had been learned, and hence the first few links had been established between the brain areas involved, a pattern was set up that, when new signs were taught, could be quickly repeated for each new sign. Only some development of this kind could account for the hundred-times-faster speed with which, after the first handful, new signs and meanings were connected.

Two things only could have triggered the growth of the neural network that made it possible to connect arbitrary signals with things in the outside world. One of them, the one that worked for apes, was a deliberate act of intervention by another species: us. The other, the one that worked for our remote ancestors, was factor X—the factor this book is looking for.

But if at least a few of the many features of language can be taught to apes—if, in other words, they can be taught some kind of protolanguage, maybe the kind our ancestors developed—how is it that they have never used that capacity for themselves, in the wild?

BACK TO THE WILD

First, we should deal with the claim "Well, they do use it in the wild— we just haven't been smart enough to understand how they do it."

This claim had a lot more going for it in the early years of primate ethology. But today, apes of several species have been studied in their natural surroundings, as well as in some zoos and research centers where those surroundings were replicated to the extent they could be. They have been studied by many acute and highly motivated observers for nearly half a century. The behaviors of these species have been described and discussed and analyzed over and over again. Yet not one researcher has ever come up with any behavior that seemed remotely languagelike. And with every year that passes, the likelihood that one ever will gets smaller and smaller.

You can't prove a negative, but if you're ever going to get anywhere in science, you just have to ignore possibilities, no matter how tempting, for which there is absolutely zero evidence.

So let's look at the opposite argument: "If they really have these capacities, how come they never use them in the wild?" (Implied: if they don't use them there, they can't have them—they must be merely artifacts of the experiments.)

Remarks like this show a serious misunderstanding of how biology works. You can make such remarks only if you believe that every potentiality existing in the genes *must* be expressed in terms of behavior. A moment's thought shows that such a belief would automatically rule out any kind of innovation. Animals would not only have to play exclusively with the deck nature dealt them—they would be compelled, like helpless automatons, to play every card in that deck. Only a mutation, a new card in their genetic deck, could bring about innovation.

There are, of course, simple creatures, programmed down to the wire, of which the foregoing might be true. Once you're past nematodes, it's a different ball game. More complex animals can learn from experience, and they couldn't do this if genetic determinism was total and absolute. When environments change, some members of a species often survive, and they can do this only by doing things that their genetic equipment allows, but that they had never done before, because they'd never *had* to do such things. Around every animal there's an envelope of potentiality, of things they're not specifically programmed to do but that they can do somehow, if they have to in order to stay alive.

In the next chapter I'll deal at much greater length with the complex relationships between genes, behavior, and environment. For now, we need only note that Washoe and Lana and Kanzi and all the other trained apes underwent a radical change in their environment, a change that, like many other environmental changes, came about through a new kind of contact with a different species—in this case, us. And like many animals before them, they successfully adapted to that change, by producing the kind of behavior—protolinguistic behavior—that seemed to be required in order to survive and prosper (read, get banana slices and M&Ms).

Which of course is very far from the behaviorist creed—gospel in the middle of the last century—that you could condition any animal to perform any kind of behavior. Eörs Szathmáry, John Maynard Smith's ex-student now with the Hungarian think tank Collegium Budapest, made a nice distinction in a recent paper between changes that are variation-limited and changes that are selection-limited. To say that some potential change is variation-limited means that the kind and/or degree of genetic variation necessary to bring that change about simply isn't present in the species concerned, and isn't likely to develop within any reasonable time frame. No matter how strong a selective pressure exists, a variation-limited change simply can't happen.

To say that a change is selection-limited, however, means that the genetic bits and pieces sufficient to bring that change about are all present (or can be very easily attained) and all that is required is a strong-enough selective pressure.

For most species, changes that might lead in the direction of language are variation-limited; the stuff that would build those changes just isn't there. For probably all four varieties of great ape—chimpanzee, bonobo, gorilla, orangutan, and hence necessarily for our own ancestors—such changes are selection-limited, awaiting only the right kind of pressure to bring them to birth.

Which immediately raises the question, are the great apes unique? Or are there other species for which the first steps toward language are only selection-limited?

Who's ready for protolanguage?

Sea lions wouldn't be on many people's list of the most intelligent species. But from the late sixties on, Ron Schusterman, working out of California State and later the University of California, Santa Cruz, showed that while sea lions weren't adapted for any productive skills, they could master most if not all of the receptive skills that apes could. Simultaneously, Lou Herman at my own institution, the University of Hawaii, was performing similar experiments with dolphins. And, perhaps most amazing of all, Irene Pepperberg, first at the University of Arizona and now at Brandeis, trained an African gray parrot, Alex, to do all the things that sea lions, dolphins, and apes had been trained to do.

Alex actually talked (I use the past tense because, unfortunately, he died in September 2007). Sure, you say, so what? Lots of parrots talk. But this one talked sense. If you asked him what he wanted, he'd tell you, "I want nut!" Give him a grape instead and he'd say indignantly, "No! I want nut!" He knew and could appropriately use some fifty words, counted up to six, identified seven colors, could do match-to-sample tests (he knew what "same" and "different" meant), and could pick, from an array of objects, the one that is green and three-cornered. Griffin, a younger parrot, has a productive capacity perhaps even equiv-alent to Kanzi's: "want grape," "go chair," "green birdie," "go back chair" (as with the apes, his combinations were spontaneous and untrained, but like theirs, they were limited to two or three units).

Are these just tricks, artifacts of clever animal training? Skeptics who insist that they are totally miss the point. *It wouldn't matter even if they all were.* You can't train animals to do things that their neural infrastructure won't allow them to do. If their neural infrastructure allows them to do those things, it can only be because that's the way the animal's genes built its neural infrastructure. In other words, if the things an animal can be trained to do amount to a form of protolanguage, then that animal is only selection-limited—it already has the range of genetic variation on which a strong-enough pressure could work to produce protolanguage.

So Irene has posed the jaw-dropping question: Must human language seek its precursors exclusively in the behavior of apes?

Most people have simply assumed that it must—that language must be a homologous rather than a merely analogous trait.

Whenever a biologist finds a trait that's shared by two or more species, his first thought is likely to be, is this a homology or an analogy? A homology is some trait or feature that the species share as a result of their common ancestry. Sometimes an ancestor with a similar feature is known to exist, but even if not, it's a fair assumption that anything shared by closely related species comes directly from their genetic inheritance. However, similar features are sometimes found in animals with only very remote genetic connections—animals that have many closer relatives lacking the feature in question. Here, the explanation is likelier to be an analogy—the shared feature represents a similar response to a particular environment. The classic example is that of dolphins, sharks, and ichthyosaurs, far from one another in the great bush of life, but each growing a similar dorsal fin in response to the need for rapid maneuvering in deep water.

Homology is commoner than analogy. Evolution seldom throws stuff away. It works, in Darwin's phrase, through "descent by modification," so any feature of a common ancestor is likely (not certain, by any means!) to show up in some form or other in species that descend from that ancestor. But analogy can never be ruled out, least of all where the same capacity shows up in species that are only very distantly related. Like the capacity for learning a protolanguage, shared not just by the more advanced primates but by dolphins, sea lions, and parrots.

Analogies represent similar solutions to similar problems. Bear this in mind; it will become very important later on.

If readiness for protolanguage came from a homology, and if that readiness had a range that included both apes and parrots, we'd have to look for a common ancestor around 300 million years ago. And we would then have to explain why so few of the countless millions of descendants of that ancestor have shown any proto-protolanguage capacities.

If that's indeed a fact, not just an artifact.

Because, to the best of my knowledge, nobody has seriously attempted to "teach language" to more than seven species—the four great apes, Californian sea lions, Atlantic bottlenose dolphins, and African gray parrots. I did once, years ago, read a news item about an Italian countess who taught her dog to type; one day, after the mutt crapped on her carpet, he went to his typewriter and tapped out "Bad dog!" (or more likely "*Cane cattivo*"). But I suspect that story was the brainchild of some cheeky cub reporter on a slow news day (if, however, it's true, please contact me immediately!).

Seriously, though, nobody knows how far down the phyla one could go and still get the kind of results obtained from our seven species. But whatever the answer, the odds are that we have an analogy rather than a homology here, and the trigger this time is not a particular type of physical environment—it's the attainment of a certain level of cognitive capacity.

It's tempting at this point to talk about "higher" animals, but I'll resist the temptation. In his diary Darwin wrote to himself the admonition: never say "higher" or "lower"; what that usually turns out to mean is more, or less, like humans. However, one can hardly refrain from talking about "more" or "less" complex: here we have objective measures, numbers of genes, numbers of cells in the brain. But what exactly do we mean, in this case, by "cognitive capacity"?

I don't actually see the level of cognition required for protolanguage being particularly high, at least not in the sense of some elaborate and convoluted mental structure. I think it has more to do with having the aural and/or visual capacity to sense the world as divisible into a large number of separate and distinct categories, and the brain space to sort and systematically store all the features that distinguish the various categories. (Note how careful I am to say that they have categories, not concepts—the distinction between these, one a lot of people don't even

bother to make, will turn out to be crucial in chapter 10.) Once such categories are in place, linking them to signals is always a possibility. As
I said at the beginning of this chapter—but it's high time for a
reminder—such signals are not true words, though they share referential properties with words. It's far from clear that these signals, for the
animals concerned, need to be truly symbolic. Such signals are likeliest
to be used, like ACS calls, when their referents are physically present. I
don't think that matters, though. I think we'll find, when we come to
human ancestors, that words got going in a very similar way, and indeed
had to get going in a very similar way, since you can't leap from icons
and indices to symbols in a single bound.

It's certainly a fact that vision is well developed in apes and parrots,
and hearing in dolphins (I'm not so sure about sea lions). Whether that's
a necessary or a sufficient precondition for protolanguage or just an
accident we won't know until more people take more species and subject them to the same training that the Magnificent Seven experienced.
And I think that it should be done. Go down to gibbons, even
macaques. Try prairie dogs to check if brain size really has anything to
do with it. Or crows, to see if the parrot's the only bird that can speak.
And we should have some way of brain-scanning all of these animals,
including the original seven, while they perform, to see if what happens bears any relation to what happens in the brain when humans use
either a natural language or a pidgin. (That will be tricky when we
get to birds—their brains are configured quite differently from those of
mammals.)

Once we stop asking the stupid question "Can animals acquire language?" (the short but unhelpful answer is *no*!) and start asking the sensible question, "What kind(s) of neural substrate are both necessary and
sufficient for protolinguistic behavior?," we can open up whole new
fields of research and start to get somewhere.

It's already pretty clear, though, that the prerequisites for something
approaching language are by no means limited to our immediate relatives. And if, under instruction, a whole range of species can learn some
kind of protolanguage, this suggests that, in any species within that
range, protolanguage is selection-limited, not variation-limited. In other
words, no special changes, magic mutations, "language organs," or dedicated circuits were needed for language to start.

Just a large enough brain, a wide enough range of categories, and, most important of all: the right selective pressure.

Pro bonobo publico

Since we now know we don't need to be descended from apes to have gotten language, and since there's no reason to think our ape heritage, other than giving us protolanguage-readiness and a highly social brain, necessarily contributed much to our getting it, we can bid farewell to our cousins and go looking for language in what, at first sight, might seem all the wrong places.

But before we do that, there's one personal experience I'd like to share with you, one that for me shed considerable light on why language didn't evolve in any species other than ours.

Not so long ago, Sue Savage-Rumbaugh kindly invited me to visit her spanking new, ten-million-dollar ape research facility, located a few miles south of Des Moines. I met with Kanzi, face to face. Right away he struck me as a personality to be reckoned with. He exuded the kind of serene confidence in his own entitlement that you seldom find outside pop stars, politicians, and the very rich. Under that was a sharp intelligence, both wise and crafty. If you met him at a scientific conference, you'd watch your arguments (or in an inner city, your wallet).

Kanzi ruled like a pasha over a bunch of bonobos; what he wanted, he got, whether it was the tastiest tidbit, the most desirable female, or the attention of his keepers. The strength of the bond between him and Sue was unmistakable. But what did it spring from? Genuine affection, the Stockholm syndrome, or some mixture of these?

For after all, Kanzi and his cohorts were prisoners. Granted, their jail was the opposite extreme to the pharmaceutical Abu Ghraibs in which some less fortunate animals languish; their keepers loved them, and all sorts of diversions were laid on for them. But they were still prisoners, not free to go when and where they chose, under the absolute control of another species. (Imagine how you'd feel if aliens from space housed you in one of their "facilities" and "studied" you, however amiably they did it.)

Under those circumstances the bonobos did just what you'd expect them to—what slaves did back on the old plantation, what any of us would do under similar conditions. They shuffled and sang, shucked and jived—yes massa, sure massa, I'll say anything you want me to say. For the rest of the time they just got on with their bonobo social lives.

Recall the conventional wisdom of today: language arose through social intelligence, to deal with the ever-growing complexity and sophistication of primate lives. To the best of my knowledge, not one of the dozens or by now probably hundreds who have endorsed this notion has ever provided a concrete example of one specific problem in social life that you can't solve without language but can solve with it. But absence of evidence seldom slows the spread of fashionable ideas.

In fact, the bonobo—according to Frans de Waal, who has studied them as closely as anyone—already "shows an unparalleled social organization" and devotes at least as much time and energy to its social interactions as any other primate, employing a "mental capacity" that "has the power of revolutionizing social relationships." I'm prepared to bet that the social life of bonobos is at least as rich and complex as the social life of the human ancestors who made the first breakthrough into language, and quite likely more so—we'll see why in chapter 6. And here bonobos were, being given language for free, so to speak. If the language-from-social-intelligence theory was correct, you'd expect them to seize on it eagerly, exploit it in their daily exchanges. Indeed, you might be baffled to explain why, given so high a level of social intelligence, they hadn't already discovered language for themselves.

I watched them over a long weekend, and they showed no sign of any real interest in language. They'd cooperate when Sue asked them to. They'd press one or two of the three hundred or so lexigrams—a set of arbitrary, abstract symbols standing for human words—on a computer screen, and a voice synthesizer would sound out the equivalent English. Then, as soon as they decently could, they'd get back to their games.

My most vivid memory of my visit concerns the plastic sheets. To encourage the bonobos to learn and use lexigrams, Sue had had them printed on sheets with thick, transparent plastic covers, about the size of a Rand McNally atlas or a fancy restaurant menu, and these sheets were scattered here and there in the apes' enclosures so they would be available any time a bonobo felt like using them spontaneously. Which was hardly ever. Far from treating them as keys to a new, richer world, the bonobos totally ignored them, except when cajoled by their keepers. For the rest of the time, these plastic dictionaries, piss-stained, fouled with the dirt that accumulates, no matter how often and diligently you clean up, wherever animals (including us) are trapped in confined spaces, were trampled underfoot—kicked around, as the Irish say, like snuff at a wake. The bonobos didn't want them, didn't need them. Bonobos just want to have fun. All the lexigrams did was get in their way.

You and I have a purely species-specific view of language. To us it's the ultimate adaptation, the core of what we are. We can't imagine a species that wouldn't be delighted to have it—that once it got it, wouldn't cling to it and exploit it as thoroughly as it could.

But that's just because we're us, because we can't imagine a life without language. For that matter, I even find it hard nowadays to imagine a life without computers, without e-mail, without word processing software. Yet I lived that life, pecked away at my Olivetti with my Wite-Out by my side, tore up and laboriously rewrote draft after draft, handwrote personal correspondence and schlepped it to the P.O., never for one moment imagining or even wishing that some whiz kid would sell me some electronic dingus that would put paid to all that labor-intensive stuff. It was just the way things were.

Other species have accepted the way things were, have lived without language and prospered from time immemorial. Bonobos handle their complex social lives quite happily without it, so why wouldn't our ancestors have done the same?

Well, because they took the road less traveled by, and that made all the difference. They opened up an ecological niche that no animal of their size and complexity had ever entered before.

5

NICHES AREN'T EVERYTHING (THEY'RE THE ONLY THING)

THE THEORY OF NICHE CONSTRUCTION

I guess there are still some people who think our ancestors got cleverer and cleverer until one fine day they just up and invented language, right off the top of their clever little heads.

But if one species and one only has a host of complex and highly developed languages and no other species, unaided, has anything you could call language at all, then language must somehow form part of the biology of our species, just as much as walking upright does. Exactly how it's embedded in human biology . . . Well, that's the question we're all trying to answer. But no serious scholar nowadays doubts that language is, at bottom, biological rather than cultural, and therefore was not created, but somehow evolved.

Theodosius Dobzhansky said it best: "Nothing in biology makes sense except in the light of evolution." But exactly what kind of evolution?

A decade ago almost everyone would have said, "What a stupid question! There's only one kind." That kind was the neo-Darwinian consensus, epitomized in the words of another icon, George Williams: "Adaptation is always asymmetrical; organisms adapt to their environment, never vice versa." True, some biologists favored a more nuanced version, but for the majority, the organism was impotent, the environment all-powerful, and any interaction between the two ran along a strictly one-way street.

All that is changing now under the impact of what is known as niche construction theory—a theory that gives animals themselves a vital role to play in their own evolution. Among its many virtues, this theory can explain both the rapid cascades of change that gave rise to Stephen Jay Gould's theory of punctuated equilibrium, and the emergence, from time to time, of things that look at first like total novelties (language is only one of many examples).

If we're going to learn about niche construction theory, there's no better place to start than with beavers.

The beaver's tale (tail?)

Everyone knows about beavers.

Beavers build their homes, known as "lodges," in places where no sudden rise in water level will sweep them away, and where no sudden fall will expose them and their kits to predators. Their favorite environment is therefore a marsh or pond. If there isn't one there already, they make one. They dam fast-running streams by chewing through stems of saplings and bushes, piling the resulting brushwood in the path of the current, and patching the gaps with mud. Sooner or later the dam holds, the water backs up, the land behind the dam is flooded.

Beavers are what ecologists call a keystone species. They create wetlands (not, alas, as quickly as we are destroying them) that serve as homes and breeding grounds for an immense variety of species—fish, crustaceans, algae, waterfowl. Beavers are a keystone species because, if you took beavers away, many other species would collapse, just as an arch collapses if you knock the keystone out. In making an environment for themselves, they made one for others too. But that's not all they made. They made themselves.

One of the things that's struck everyone interested in nature is the way in which species fit into their particular habitats and ways of life as precisely as a key fits a lock. (If you've ever had a new key cut and found that the slightest roughness made it unusable, you know what I mean.) Beavers fit their habitat in just this way. Their teeth are massive, chisel-like objects, just fine for ripping through the toughest bark. Their mouths are shaped so that, if the stems they are chewing through are underwater, the water doesn't get into their throats and choke them. Glands under their skin pump out oils that effectively waterproof their thick fur. Their feet are webbed for stronger swimming, their lungs disproportionately large so they can work underwater for long periods, their eyelids transparent to protect their eyes while still allowing them to see clearly enough beneath the surface. And their tails are long and flat, driving and steering them in the water, radiating heat when they cross dry land on hot days, storing fat against lean seasons.

Surely, you think, someone or something must have designed them specifically for their job.

No, says the orthodox evolutionary biologist. Genetic variations occur in any species, all the time. The environment simply selects from those variations the ones that best fit current conditions. If the world cools, the animals with thicker fur survive better and their offspring outnumber the offspring of those with thinner fur, so eventually everyone gets thick fur. If the world starts warming again, same thing happens, only in reverse this time.

Sure, you say, but we're not talking about things like responding to climate change. We're talking about very task-specific adaptations to very specialized ways of life. Don't tell me those come from just chance variation.

I don't, says the evolutionary biologist. It's not just chance. It's chance *plus necessity*! The necessities of the environment, all the things you have to cope with to survive. These are the things that select from the variations, and that's all we mean when we talk about "natural selection."

But wouldn't that take an awful long . . . ?

Time? Of course, says the evolutionary biologist. But that's no problem. Evolution has oodles of time! All the time in the world! In fact:

Chance + Necessity + Time = Perfect Fitness. QED.

It's hardly surprising that a lot of people, people who are by no means the knuckle-walking retards of evolutionist fantasies, have felt dissatisfied with this. Many have been driven into creationism or Intelligent Design precisely because they felt there was something missing from the equation—that to accept it at face value meant as big a leap of faith as believing in some all-powerful agency.

THE EVOLUTION OF EVOLUTION

Think back for a moment to Jean-Baptiste Lamarck, the French naturalist who wrote about evolution fifty years ahead of Darwin. So why do we talk about Darwinism and not Lamarckism? Because Lamarck put all his money on a major evolutionary mechanism that seemed to explain why animals so neatly fitted their environments; only problem was, it turned out to be wrong—at least, wrong in the form in which he stated it. Darwin, on the other hand, hedged his bets; it's hard to think of any evolutionary theory that you couldn't support from some quote somewhere in his voluminous writings, including even Lamarckism. So Lamarck, though in many ways ahead of his time, finally lost out.

In the nineteenth century, many people suspected that evolution happened, but nobody knew what made it happen. Lamarck claimed that it happened because acquired characteristics could be inherited— the results of things animals did in their lifetimes could be passed on to their descendants. If they used some part of their bodies more than others, that part would grow and become stronger in their offspring. If they adopted and practiced some new behavior, if say an originally short-necked animal started reaching for the leaves on higher branches, then its children and its children's children would grow longer and longer necks, and sooner or later, lo and behold, you got giraffes.

When the Austrian monk Gregor Mendel grew sweet peas and showed that what caused them to vary in color from generation to generation had to be something inside the seed, rather than what the plant itself did, the scientific silence was deafening, and Lamarck's theory still had devoted followers. It wasn't until early last century that researchers put Darwin and Mendel together, the science of genetics emerged from their union, and the rest, as they say, is history.

But poor old Lamarck and his ideas became History with a big H. How could anyone go on believing that what animals did had any effect on their evolution, once we knew that genes did the trick?

Genes aren't the only thing, of course, though you could be forgiven for thinking so. In the neo-Darwinian consensus that has dominated biology for a century, animals are just vehicles for their genes. They mate, they breed, they fight for survival, but apart from that they don't *do* much; they merely exist as a source of genetic variation from which natural selection can select what is most appropriate under current conditions. And those conditions, of course, are almost always changing. Active environment, hyperactive genes, passive animals—that was the picture neo-Darwinism painted.

In all of this, a vital factor was overlooked.

From time to time, new behaviors appear. (If they didn't, we would all still be swimming around in the primordial soup.) Where do those behaviors come from? Do the genetic changes come first, then the new behaviors? Certainly not all the time. More often than not, behavior changes first, then the genes change to keep pace with it.

Take weaning, for example—the process by which all infant mammals switch from mother's milk to their regular species diet (whatever that is). Since there's no animal Doctor Spock to tell other species how to rear their children, nature has to take care of it. Nature does this by making it so that, at some point in their infancy, mammals cease to be able to digest milk. It's not a case of fight or switch; it's a case of switch or starve. We can understand how this came about by thinking what would have happened if it hadn't.

For other mammals, the only source of milk is mother—they have no domesticated animals to serve as an alternative source. Fine so far, because mother's milk is one of the most nutritious substances known. If children could keep on drinking it, they surely would. But if they didn't stop, two things would happen. First, mothers would quickly get exhausted. As things are, weaning gives them a break between bouts of nursing, enabling them to recoup both their energy and their milk supply. But a constantly exhausted mother can't give her children all the attention they need, and the children would suffer.

Second, very soon there would be two infants (or two sets of infants, for species with multiple births) competing for milk, then three. Very soon the mother would find herself physically incapable of providing sufficient milk for all of them, and in consequence all would suffer from an inadequate diet. Therefore, if there were children who were lactose tolerant beyond the normal age for weaning (that is, children who could digest lactose, the sugar that makes milk indigestible to adults), these would grow up undernourished and unhealthy compared to those who quit the breast early. Even if they didn't starve to death in childhood they'd have shorter lives and leave fewer children of their own. Thus, over time, lactose intolerant animals would win out and eventually become universal. That's natural selection at work.

But now for some unnatural selection. If you can drink milk without getting stomach problems, and there are still many in our species who can't, then you have herdsmen (and herdswomen) somewhere in your family tree. Several thousand years ago some of our ancestors started domesticating animals. In many places they did this where there was little to eat apart from what those animals produced. Their milk, for example; it formed too rich a food source to neglect, and one that, however much you used it, would put no strain at all on your mother.

So people tried drinking it, and it turned out there was a rare mutation on chromosome 2 that in a tiny handful of adults prevented lactose intolerance from developing. (Note that before domestication, this mutation, like most mutations, would have been actively dysfunctional.) These adults drank milk and it made them healthier, just like the milk advertisements say. It gave them a fractional advantage in spreading their genes at the expense of their lactose-intolerant cousins. And, given a few thousand years, a fractional advantage is all it takes for your genes to spread and prosper. Today, 98 percent of Swedes and 88 percent of white Americans are lactose tolerant, while Chinese and Native Americans are respectively 7 and 0 percent tolerant. (Few Chinese and no Native Americans have herding ancestors.)

Now if none of our ancestors had gone for the herding lifestyle, few if any of today's adults would be able to drink milk. In other words, here's a genetic change resulting directly from a new thing that people themselves chose to do. Of course what they chose to do didn't *cause* the mutation, the evolutionary biologist will remind you—that was pure genetic accident. Perhaps, but the fact remains that, if we hadn't domesticated animals, that mutation would have been actively deleterious and would probably have been bred out of the species. Here and in this way, at the very least, humans have taken a hand in their own evolution.

Lamarck had been wrong in his choice of mechanism; genes, not lifetime achievements, are the motor that drives evolution. But his intuition that animals themselves guided their own evolution was dead on target. Because it's the interaction of genes and behavior that starts the evolutionary motor, and the feedback between genes and behavior that keeps it going. That's the insight that gave birth to niche construction theory.

How niche construction theory changed my life

Barcelona, summer of 2004. I was there for two consecutive gigs: first a weeklong symposium on the evolution of cognition, and then to participate in Barcelona Forum 2004, a mammoth cultural funfair intended to establish the city as the new Paris, the intellectual capital of Western Europe.

One of the other speakers in the symposium was the philosopher Daniel Dennett, probably best known to you from his book *Darwin's Dangerous Idea*; we'd last met in Budapest two years before, where we fought for a month over memes (he for them, me against). Another speaker I didn't know: Marcus Feldman, an Australian working out of Stanford, who was talking on niche construction. I'd never heard of it.

"Do you know this stuff?" I asked Dennett.

"Sure," he replied. Dan is one of those infuriating people who always hear about everything before you do. "It's important, too," he added.

"Come on," I said. "Everyone knows about beavers."

"No," Dennett said, "there's a lot more to it than that. You should listen up and listen up good."

I stayed skeptical. Nine new things out of ten in science turn out to be passing fancies, so "skeptical" should be one's default setting. But apart from his inexplicable infatuation with memes, I have the highest respect for Dan; out of all philosophers, he keeps most on top of every new development in computer science and evolutionary studies.

Came the forum. I was on a panel debating what science could do for world peace. Nothing, I said, except to point out that humans, far from being the gentrified primates that evolutionary psychologists say we are, have chosen a lifestyle much closer to that of ants, and the tensions between our ape nature and our ant circumstances create problems that are probably insoluble. It was actually a niche construction talk, or rather one that badly needed niche construction theory to give it validity and coherence. Of course I didn't realize that at the time.

On the last day of the panel John Odling-Smee spoke. John, who's at Oxford University, is, like Feldman, one of the three cofounders of niche construction theory. (The third is Kevin Laland of the University of St. Andrews.) I felt my resistance crumbling. When John finished I asked him a very loaded question, which he answered with grace and skill. We got to talking afterward, and went to lunch together in the hotel we were both staying at, a neo-Japanese monstrosity where the black-robed waiters looked like Buddhist monks. I was starting to get really excited, because I could now see that with this theory a lot of vague ideas that had been floating around my head for a long time might come together and make sense.

Over the next few weeks, John very kindly e-mailed me a slew of articles on niche construction, some for it, some against, and I eagerly read the first and so far the only book on it, *Niche Construction: The Neglected Process in Evolution*, by Odling-Smee, Laland, and Feldman, which had come out the previous year. And I became hooked. This was one of those ideas that are so beautiful, so stunningly simple, you wonder why nobody ever thought of it before.

What the theory says

Well, that's not quite fair. Richard Lewontin, Conrad Waddington, and other biologists had emphasized the importance of behavior in evolution. Richard Dawkins, who became one of niche construction's sharpest critics, wrote a book called *The Extended Phenotype* that anticipated certain aspects of the theory. Dawkins wanted to modify the concept of phenotype—formerly just the expression of an animal's genes in terms of its own body shape and skills—to include the artifacts, if any, that the animal constructed. In other words, a beaver's dam was just as much an expression of beaver genes as a beaver's tail. But Dawkins's approach still focused exclusively on genes and what genes did; in his own words, "an animal's behaviour tends to maximize the survival of the genes 'for' that behaviour." According to the niche constructionists, this was only the beginning of the story.

The basic idea to hold in mind is that animals themselves modify the environments they live in, and that these modified environments, in turn, select for further genetic variations in the animal. So a feedback process begins, a two-way street in which the animal is developing the niche and the niche is developing the animal, until you get the lock-and-key fit between animal and niche that makes people say, "But there *must* be a designer!" Animals aren't just passive vehicles for their genes; they play an active role in designing their own destiny.

So what exactly is a niche, anyway?

According to Eugene Odum, author of *Fundamentals of Ecology*, "the ecological niche of an organism depends not only on where it lives but also on what it does. By analogy, it may be said that the habitat is the organism's 'address,' and the niche is its 'profession,' biologically speaking." In fact we can distinguish not just two but three essential components of a niche:

- Habitat: a particular type of environment that can be macro (savanna, rain forest, marsh, mountain, tundra . . .) and/ or micro (topsoil, tree bark, pond scum, nest, burrow, mound . . .).
- Nourishment: a particular type of food (grass, meat, insects, honey, microorganisms, fruit, blood . . . or some combination of these and/or other things).
- Means: a particular way of obtaining that food (foraging, scavenging, stalking, pack hunting, ambushing, sieving, digging . . .).

Thus the hyena niche consists in living on open savannas, eating meat, and scavenging or hunting in packs. The niche for baleen whales consists in living in the open ocean and eating marine microorganisms obtained by sieving seawater. The frog niche consists in living in ponds or swamps and eating insects for which the frog lies in wait.

People haven't often thought of niches as being actively constructed. They've tended to think of them as ready-made, just waiting for some animal to step into them, and have regarded beavers, for example, as quaint exceptions to this rule. Not so; the authors of *Niche Construction* list hundreds of species that, to some degree or other, engineer their own niches. Beavers, ant species like leaf-cutters that build underground fungus farms, and earthworms are just some of the more dramatic cases.

Take earthworms, for instance.

Darwin loved earthworms. He studied them perhaps more than any other organism. His last published book was about worms. If worms had given him the same depth of insight that Galapagos finches did, evolutionary biology might have developed very differently. But they couldn't have. To understand what makes earthworms special, you have to already know something about modern, gene-centered versions of evolutionary theory.

They're called earthworms, but they weren't always. They began life as waterworms. But just as ancestors of whales and dolphins went from land to water, so these worms made the reverse journey. Only they didn't do what whales and dolphins did, what evolutionary science says they should have done—simply and straightforwardly adapt themselves to the new environment. Instead, at least in part, they adapted the new environment to themselves.

Makes sense when you think about it. There's not much you can do with water. But earth is malleable; you can mine it, shape it, even eat it if you have the right kind of stomach. So earthworms-to-be set about changing the land.

They didn't change their kidneys. They didn't change their production of urine. (Water-dwelling organisms put out more urine than land dwellers, mostly to get rid of excess salt.) They didn't change any of their other bodily forms or functions in ways that so drastic a transition might lead you to expect. Instead, they transformed the ground itself. They started by exuding quantities of mucus that softened the ground and made it slippery so they could dig and navigate a system of tunnels. Then they dragged bits of decaying vegetation into these tunnels and mixed it with inorganic material and ate the mix. What they then excreted—known as "worm casts"—is so rich in minerals and fine in texture that avid gardeners, my wife included, keep round metal drums full of worms, feed them the household scraps, and reap the richest compost known. (What's more, the brown liquid that collects in the bottom of the drum, in which, Yvonne says, "they love to take a swim," makes a fantastic organic fertilizer.)

But wait, those are domesticated worms. Wild worms have a harder row to hoe, but they're not all that different. Stick a spade into hard compacted earth that hasn't ever grown much in the way of plants and you probably won't see any worms. Put the same spade into loose crumbly loam and you'll turn up a worm or two almost every time. You might think, oh well, that's just because I've dug and fertilized that part of the garden. Maybe, but there was rich loamy soil long before there were human gardeners, because generations of worms had worked on it, breaking it up, enriching it with their casts. It was a mutual process, niche and organism changing together—worms putting out more mucus, getting to digest more varied substances, while the earth around them grew ever richer in nutrients and easier for them to tunnel. These mutually reinforcing processes have gone on for countless worm generations, fitting worms to earth, earth to worms, and incidentally making the world a happier place for insects, plants, gardeners, and other organisms that benefit from enriched soils.

This brings us to what's become the most controversial part of niche construction theory.

Doing things that have knock-on effects for your own species is one thing. Doing things that have knock-on effects for other species, say the critics, is a horse of another color entirely. Especially if it's mostly other species that benefit, not your own.

For example, consider photosynthesis, the way in which plants use the energy supplied by sunlight to turn water and carbon dioxide into carbohydrates essential for their growth. In the process, oxygen is released as a by-product. But the chief beneficiaries of this oxygen are not plants, but other living organisms—in fact, every living organism that needs air to breathe. Before plants started, there was nowhere near enough oxygen in earth's atmosphere to support mobile, energy-hogging creatures once these got any bigger than a single cell. It was only when plants had pushed oxygen levels way higher that multicelled creatures could grow and prosper. The niche that plants had constructed—staying in one place, drawing ingredients from ground and air, and using solar energy to process them—had altered the genetic future not of themselves but of countless other species.

But that isn't evolution, some biologists protested. Evolution is about individuals that can reproduce themselves. Individuals of the same species, naturally. If the behavior of individual A in species X influences the genes of A's offspring, that's one thing. They're the same genes; you can measure the variance from one generation to another, relate it to behavior, make scientific statements, maybe even predictions. But if the behavior of individual A in species X influences the genes of individual B in species Y and *its* offspring, there's no common ground. You don't have anything objective, like a gene, to measure; you've got no physical link between cause and effect. To propose a single theory that covers both processes—the effects of niche construction on the niche-constructing species and the effects of that species' niche construction on other species—is, according to Dawkins, "pernicious."

Fortunately, we don't have to get involved in this argument, because in what follows we'll be concerned with only one species, or rather one series of species—our own and its immediate ancestors. The only genes involved are those that would eventually produce you and me. Nor do we have to struggle with the high-powered math that Odling-Smee and his colleagues use to justify their ways to population geneticists. All we really need is the take-home message:

It's not just the species that makes the niche: it's the niche that also makes the species.

Where humans fit in

What I dimly perceived in Barcelona (it took me months to figure out all the angles) was that, without niche construction theory, you'd never arrive at any satisfactory explanation of how humans got language.

For quite a while, I'd been aware that one of the greatest weaknesses in language evolution studies was a failure to integrate them into an overall account of how the human species as a whole had evolved. You could read not just papers but whole books on language evolution that had hardly a mention of what human ancestors were like or what they were doing while language evolved. It was as if the authors were writing about how species X got language on planet Y. Or worse, since even on planet Y you'd have *some* physical constraints, they seemed to be writing about how abstract beings might get language in a Platonic empyrean. Linguists weren't the only ones who did this, though they were among the worst offenders.

But this couldn't be right. Language didn't evolve in a vacuum. It was, it had to be, an evolutionary adaptation, just as much as walking upright, shedding body hair, or getting an opposable thumb. And it's no longer true to say we don't know enough about our ancestors to provide any significant input. It's true that we don't know enough, period.

Unless we can invent a time machine, or find a way to clone those ancestors from fossil DNA—just wild dreams, as things are—we never will know enough. And even if we cloned those ancestors, we couldn't clone their environment, so there'd always be questions. But we do know enough, right now, to draw broad if sketchy pictures of our past, and use that information to separate plausible from implausible proposals. It isn't facts that are lacking so much as a way of focusing those facts, putting them into a coherent perspective.

Niche construction looked like that way, not just for language but for human evolution as a whole.

Nowhere is this more apparent than in the field of human culture. Humans and their culture have always presented problems for the life sciences. Human culture, with all its multifaceted complexity, its centrality to all we think and do, seems to be the only thing of its kind in nature. And our science just doesn't know how to proceed with a population of one.

Of course there are those who claim that some animals do have cultures. If you define culture as they do, in terms of the passing on from one animal to another of some learned behavior, that's trivially true. Like the Japanese macaque monkeys, one of whom learned to wash sand off potatoes before eating them; other monkeys, observing this, followed suit, and it became a custom. Or the chimpanzees of the Ivory Coast, who break open palm nuts and show their children how to do it. But these "animal cultures" are so limited, so impoverished, compared with the immense and constantly expanding human complex of traditions and stories and arts and artifacts and sciences, that any comparison looks like a desperate attempt to carry out some kind of agenda.

As indeed it is. For millennia, religions have been telling us that humans are a special creation, something set divinely apart from the rest of nature. Now that we know that this is false, the pendulum has swung to the other extreme. Rationalists eager to prove their faith feel obliged to insist, despite massive evidence to the contrary, that humans are no different from any other species. Any trait we have, other species too must have, and if that trait isn't as developed as it is in our version—if potato-washing and nut-cracking don't quite measure up to an Einstein equation or a Beethoven sonata—well, that's about how our swimming rates alongside a dolphin's swimming, or how our ability to locate things through the heat they give off compares with the same ability in pit vipers. Every species has things it's better at than others, and who are we to decide that our best tricks have somehow more intrinsic worth than the best tricks of others?

Ironically, niche construction theory links humans with other creatures in a far broader and more valid way than any claims about chimp culture.

Human culture is simply a case of niche construction.

LEARNING VERSUS INSTINCT—DOES IT REALLY MATTER?

Once, when I knew no better, when niche construction theory was hardly even a twinkle in John Odling-Smee's eye, I wrote, under the influence of George Williams and his like, "While other species adapt to the environment, we adapt the environment to ourselves." Now I know better. I know that many, perhaps even most, other species adapt the environment to their own needs, insofar as they have the ability to do so. Some don't have very much ability. We have more than any other species, but what we're doing is basically the same as what they're doing.

We build huge climate-controlled buildings to protect us from the heat of summer and the cold of winter. So do termites. They build mounds whose size, relative to that of their builders, is proportionately far larger than the size of our skyscrapers. They regulate the temperature of their mounds in a variety of ways; by positioning them on a north-south axis (thus limiting exposure to midday heat), by building thick outer walls, by plugging entrances at night, and by incorporating a number of air-conditioning devices: ventilation chambers, cooling vanes, air ducts, and chimneys. They also add mushroom- or cone-shaped structures to their mounds to ensure that rainwater doesn't get into them.

We have developed complex forms of agriculture to supply ourselves with food far in excess of what nature alone could furnish. So have leaf-cutter ants. Leaf-cutters create huge underground farms; in making just one of these they can shift anything up to forty or fifty tons (yes, tons!) of soil. Such a farm may employ a workforce far more numerous than that of any human enterprise (except perhaps Wal-Mart or the Chinese army). These workers climb plants, cut off leaves, carry the leaves to the farm, drag them down into special chambers, chew them to pulp, arrange them in beds, impregnate them with a moldlike form of fungus, weed the beds, and then gather and distribute the mature fungus to their nestmates.

Well, you may say, ants and termites do these things by instinct; we do them by learning. If you took illiterates from the Arabian desert or the Amazon jungle, people who'd never had any contact with an advanced technological society, and trained them to do such things, they'd do them. But you can't train an ant or a termite to do anything.

True, but irrelevant. Instinct is simply behavior stamped into the genes. Ants and termites had to start doing the things they do (in some much simpler and more rudimentary way, of course) with the genes they already had, before the interaction of genes and behavior could kick in to eventually produce complex new behaviors. Just as we had to start building crude temporary shelters of stones and brushwood long before we could erect cathedrals and high-rise apartments. Just as we had to start gathering and chewing wild seeds long before we could get to cover a large chunk of the earth with our crops. It was never the case that someone said, "Hey, guys, hunting and gathering sucks—isn't it time we tried farming?"

The truth is, we gradually and quite accidentally stumbled into agriculture, and so did ants. The difference is that our capacity for learning new things and passing them on to others (a capacity based entirely on our command of language, as we'll see in the final chapters) made us able to develop far faster than any other species—we could construct our niches without having to wait on interminable rounds of feedback between genes and behavior. But apart from that, the motivation, the process of niche construction itself, and even, as we just saw, some of its specific results, are similar across species, even species as remote from one another as termites and we are.

Indeed, the habits and artifacts of bees and termites resemble human niche construction much more than does nut-cracking by chimps or potato-washing by monkeys. Accordingly, niche construction links us more firmly to the rest of nature than any claim about "animal cultures" can. Indeed, when you come to think about it, people who make such claims are simply following the homocentric bias I've referred to in earlier chapters.

Why should the capacity for cultural learning be rated so highly? Because it's something that's shared, to a very limited extent, only by those species most similar to ours and most closely related to ours. In other words, those who seek desperately for evidence of "culture" in other animals are simply looking for humanlike traits in those animals. They're judging other species by human criteria. Let's face it—vigorously though they'll deny the accusation, they're just humanists in drag.

Niche construction is neutral. It enables us to compare the activities of species, including our own, in a completely objective and impartial fashion. And moreover, the study of human niche construction will show that this same hyperdevelopment of learned behaviors is itself based on an instinct: the language instinct.

For language is a prime example of niche construction, arising out of a specific niche and enabling us to construct more and more elaborate niches. It began as a behavior that drove genetic change and continued as a series of genetic changes that drove behavior. Is language cultural or biological? It's a truism to say "both," but while scholars have fought for generations over what biology contributes and what culture contributes, few have looked at how biology and culture might have interacted with each other over time to create the kind of language we have today. That's largely because there was no kind of framework, nothing like niche construction theory, that would have allowed them to frame the issues in more productive ways.

In fact, niche construction happens to be just the right kind of framework for examining how language was born. As noted in chapter 1, we have to explain not just why we have language but why all other species don't, so we must seek the origins of language not in things we share with other species but in things that make us different. And I said at the beginning of this chapter that there must be some large but hidden difference between apes and us. That difference could hardly lie in genetic material, since ours is almost identical with theirs. It was much likelier to exist in the niche, or rather niches, that human ancestors constructed, for these were very different from the niches of all the other great apes. Therefore somewhere, in one or other of those niches, the difference that gave us language must surely lie.

In the remainder of this book, we'll review the story of our ancestors and how they came by language in the light shed by niche construction theory; we'll find that crucial difference; and in the process we'll learn new and exciting things, not just about language but about the whole process of becoming human.

6

OUR ANCESTORS
IN THEIR NICHES

WHAT A DIFFERENCE A NICHE MAKES!

What makes us different from our closest relatives, the bonobos and chimps in whose company we spent most of chapters 3 and 4, is the series of niches our ancestors developed. This series contained at least six distinct niches: a terrestrial omnivore niche, a low-end scavenging niche, a high-end scavenging niche, a hunting-and-gathering niche, a herding niche, and an agricultural niche. You could add an urban-industrial niche, if you like, to make seven.

In contrast, the bonobo-chimpanzee line may have developed no more than one new niche since the split from our last common ancestor. At least there's no evidence for more than one. On the other hand, there's ample evidence that human ancestors constructed new niches on an unprecedented scale and at an unprecedented speed. This has to be what made our fate so different from that of the other great apes.

What I've just told you isn't the story you would have gotten until very recently. Until very recently you would have been told that changing environments selected from genes, and changing climates selected for flexibility, and this combination made human ancestors smarter, so they made tools, and then better tools, because they got smarter still—end of story.

And here's what we typically learn about the mechanics of that evolution from quasi-popular sources, which merely distill and encapsulate the scholarly sources (this extract is from the MSN Encarta website, but you can find similar stuff all over the place): "Over time, genetic change can alter a species's overall way of life, such as *what it eats,* how it grows, and *where it can live.* Genetic changes can improve the ability of organisms to survive, reproduce, and, in animals, raise offspring. This process is called adaptation . . . Many factors can favor new adaptations, but changes in the environment often play a role. Ancestral human species *adapted to new environments as their genes changed,* altering their anatomy (physical body structure), physiology (bodily functions, such as digestion), and *behavior*" (my italics).

Not a word about niches, or about any proactive role the ancestors themselves might have played in all this. No suggestion that animals, including our ancestors, could have decided for themselves what they would eat and where they would live, without waiting for genetic changes to adapt them to it (whatever "it" was). No suggestion that they might have explored new environments and begun to adapt those environments to themselves with the help of genes they already had. And no hint at all as to how language began.

The gene-centered version of evolution cannot explain how language evolved. I don't mean just hasn't explained it, or can't explain it yet; I mean it can't explain language evolution *in principle.*

Despite all the recent brouhaha over the FOXP2 gene, nobody as yet has found a "gene for language," and they likely never will. The highest probability is that language nowadays results from the interaction of a set of pleiotropic genes, that is, a set of several genes each of which performs several different functions in development. One or more of those other, nonlinguistic functions was almost certainly what each "language" gene was originally selected for. After all, humans have less than twice the number of genes that nematodes have, and since nature seldom throws stuff away, many of these are still the same genes—and whatever nematodes use them for, it sure ain't language. Recent developments in biology have shown us that genes are far more flexible than was originally thought, able to change their expression to yield a variety of results. And in previous chapters we looked at some very good reasons why the beginning of language need not have involved anything in the way of genetic change.

Language is a living proof of niche construction theory. We'll see why as we proceed to rewrite our own prehistory from a niche-centered perspective.

In the beginning

We don't, unfortunately, know what was the last common ancestor of us and the chimps. More to the point, we don't even know what that ancestor was like.

A natural assumption is that it was more or less like a contemporary ape. That would give us a simple story: we changed, apes didn't. It's plausible, but there's really no evidence for what the ancestor looked like or how it lived. And we know that there are sharp behavioral differences between chimpanzees and bonobos, which separated only a couple of million years ago.

Chimps are aggressive; bonobos are peaceful. Chimps often move in all-male groups, bonobos almost always in mixed groups. Chimps commit infanticide; bonobos don't. Chimps use tools; bonobos, for all their intelligence, don't. Female chimps are sexually available only in estrus; female bonobos are available any time. Bonobos favor the missionary position; chimps do it doggy style. And so on.

All these differences may come from a very simple niche distinction—the one I referred to at the beginning of this chapter, maybe the only new one in the whole chimp-bonobo history. Where bonobos live, edible plants abound on the ground, and bonobos snack on them while doing their serious foraging. Where chimps live, there are no plants of this kind. (If you want to know how all these consequences could follow from such a simple and seemingly trivial fact, read work by Frans de Waal or Richard Wrangham—it's fascinating, though it's got nothing to do with language.)

So was our last common ancestor more like a chimpanzee or more like a bonobo? You will actually see this debated on political lines—conservatives for chimps, liberals for bonobos. That's ridiculous, of course: first because you can't cherry-pick your ancestors, and second because the last common ancestor could have been at least as different from both chimps and bonobos as they are from each other.

At the same time, there are deep behavioral similarities between chimps and bonobos, most of which sharply differentiate them from our ancestors as well as from us. They both mostly inhabit deep tropical rain forests (though some chimps seem to thrive on the fringes). They both live mainly on fruit and nuts. They hunt and eat meat sporadically (chimps more than bonobos). They apparently never scavenge. And, despite their intense sociality and their talents when trained, they have no trace of language.

In the days of the common ancestor, forests stretched across central Africa from coast to coast, and the ancestral species, we may assume, roamed far and wide. Then seven or eight million years ago the climate began to change; paleoclimatologists have several theories why, but they needn't concern us here. The Congo basin and the rest of lowland West Africa remained mostly wet and forested, but the eastern side of the continent grew progressively drier. It did not, as some outdated accounts suggest, turn into savanna overnight. The drying trend (interrupted by pluvials, long phases of increased rainfall) extended over millions of years. Unbroken forest gradually gave way to a mosaic of forest remnants, scattered woodlands, and grass. And in response to the loss of Eden came species that could survive in this strange new scenery—australopithecines, the "southern apes" (so called not because there was anything especially southern about them, but because the first one to be discovered, by a South African, Raymond Dart, happened to lie on his home turf).

Australopithecines came in two brands, each with assorted flavors, and it's unlikely we've yet heard the end of the arguments about who they were, how many of them there were, which was which, and which, if any, gave birth to the first species that we dignify with the prefix *Homo*. (In idle moments I like to wonder whether an alien paleontologist from some species quite unlike ours would analyze and divide the fossil bones the same way we do—a fruitless exercise in itself, but it helps remind us of the inevitable subjectivity latent in any research that we conduct on ourselves.)

The two brands are known as gracile and robust—the gracile were relatively slight and skinny, the robust stockier. One of the few things paleontologists agree on is that the robust ones had nothing to do with us. They had big teeth adapted for eating uncooked tubers, and presumably they went around with digging sticks digging them up. The last of them died out around a million years ago. Maybe they ran out of tubers, maybe our ancestors ate them—who knows?

It's only the others, the graciles, that will concern us here. One at least was probably ancestral to us. Some if not all of them walked on two legs, and their (more or less) erect bipedal stance set in motion physiological changes that would come in handy when language began. Other than that, they didn't differ that much from their ape ancestors. Their brains were little if any bigger. They didn't make stone tools, at least not until the last one, *Australopithecus garhi*, and we'll see why in a moment. Their ACSs probably didn't differ that much from those of apes, except they very likely added predator warning calls.

New dangers and their consequences

Why? Let's look at the gracile australopithecine niche. A good deal of this is inference, but given what we know of the climate, the terrain, and their physiology, they didn't have a wide range of options. In the mosaic woodland they inhabited, fruit would be scarcer than in the forests, so they'd have to become more omnivorous than apes. Marks of wear on their teeth suggest that, though they didn't get to growing the huge molars of their robust cousins, they relied on roots to see them through tough times. Doubtless they took advantage of things like birds' eggs, small lizards, even caterpillars. (This may trigger your yuck factor, but modern hunter-gatherers still regard caterpillars as delicacies—in South America, an Akawaio tribesman once offered me one, several inches long, yellowish green in color, and covered with hairy warts, and he looked quite hurt when I refused it.) Almost certainly they opportunistically hunted small mammals. What is quite certain is they didn't go after the big ones. The big ones went after them.

Something many people don't care to think about is that our early ancestors were prey more often than predators.

In fact, there's a recent book—its title taking off from all that macho "man the hunter" stuff that was popular in the 1970s—called *Man the Hunted*. Unfortunately, it's spoiled by the political correctness of its stance (the authors think humans are sweet peace-loving folk at heart), but much of what it has to say about the perils of Pliocene and Pleistocene is all too true.

Australopithecines were small—around four feet high, weighing a hundred pounds or less even soaking wet. In mixed woodland and savanna country they were exposed to a wide range of predators larger and more fearsome than those of today. There were half a dozen genera of big cats—*genera*, not species; each genera contained several species. Their very names are enough to induce fear: *Vampyrictis, Machairodus, Dinofelis, Megantereon*. There was a hyena, *Percrocuta*, as big as a small lion. There was a giant weasel, *Ekorus*, nearly two feet high at the shoulder; since it could chase down pigs and small horses, it's more than likely that some of our ancestors suffered the ignominious fate of being eaten by weasels. Some of them, it's pretty certain, were killed by birds.

One of these australopithecines was the first ever to be discovered— the Taung child, a three-year-old that died around 2.5 million years ago. The back of its skull shows the keyhole-shaped incision that's typical of an eagle strike. Its eye sockets are chipped and scratched where the eagle tore out those delicious morsels, its eyeballs. (Hopefully it was no longer alive by then.)

Think of scenes like this, then think of all those currently popular scenarios about how our ancestors' social lives just kept getting richer and richer until they had to develop language in order to cope. Not one of the numerous statements of this thesis that I've read contains a single word about the ecological context in which this "increasingly complex" social life had to be lived. Obviously these authors used the ape model—just draw a straight line directly from apes to us and imagine our forebears cruising along it, sharpening their social intelligence as they went, with never a glitch in their busy social lives.

A reality check quickly disrupts this picture. Under the heavy pre-dation australopithecines underwent, in a terrain where food was scattered and hard to come by, far too much time would have been spent watching out for and avoiding predators and struggling to get enough to eat. They simply hadn't either the time or the security for lollygagging around like contemporary apes, full-bellied and relaxed, schmoozing and politicking away for all they're worth. The complex "Machiavellian strategies," the constant trying to get the better of one another that's so often seen as the locomotive of human intelligence and language, would have actively interfered with their primary goals: food and survival.

What would have been the real consequences of the australopith-ecine niche?

An increase in reciprocity among both kin and nonkin is one likely consequence. When you are at constant risk from both land and air (not to mention rivers full of ravenous reptiles), you want to make sure your buddy is watching your back, and the best way to do that is by watching his. This is the model we need to keep in mind, rather than the constant outwitting of one another, the steady escalation of mind-reading levels ("I know that he knows that I know . . .") that Steven Pinker and others see as somehow leading to language.

The only sense in which australopithecine social life would have been richer than ape social life lies precisely in the muting of within-group competition (and, ultimately, the birth of cooperation) that follows inevitably when you have to compete with other species more than with other members of your own species. Without the abundant free time and freedom from interference that easy food-gathering and relative absence of predation gave chimps and bonobos, social life would lose much of its complexity. Group cohesiveness would increase at the expense of competition. And this is not, please note, an argument for group selection. Every australopithecine was still doing its own business, serving and saving its own genes. But to do that required, as an absolute minimum, staying alive to do it, and only cooperation with other group members—whether in giving warnings or evading pursuit or resisting attacks—could ensure that they stayed alive.

Australopithecines, related even more closely than we to the great-ape family, surely had genes that would have made possible complex competitive social behaviors, if there'd been the time and security for such things. But people often forget that, except perhaps for the very simplest creatures, genes do not mandate behaviors. They simply make them possible. Circumstances will determine how far, if at all, those possibilities are realized.

For when genes and environment pull in opposite directions, environment wins. It has to. It makes sure that those who don't obey its demands die, and their genes die with them. In those who survive, the genes that can support elaborate social behaviors are simply expressed differently, or suppressed.

LIKE A VERVET (BUT NOT FOR THE VERY FIRST TIME!)

In fact, the situation australopithecines found themselves in had a lot in common with the situation of vervet monkeys today.

I noted a couple of chapters back the fact, puzzling to some, that while the relatively unsmart vervets had alarm calls for several predator species—things with "functional reference" that some even took as "precursors of words"—the much smarter chimps and bonobos had none. This makes no sense at all if you subscribe to the ladder of language notion, with intelligence and language precursors going hand-in-hand up the ladder. But it makes perfect sense from a niche construction perspective. From that perspective, you make whatever you need, regardless of whether you're a "higher" or "lower" species. Vervets needed alarm calls because they were heavily predated. Chimps and bonobos didn't, because they weren't.

Vervets now occupy pretty much the kind of terrain that australopithecines occupied then—a mosaic of woods and grasslands. Vervets now suffer the same kind of predation, aerial as well as terrestrial, as australopithecines suffered then. So it's more than likely that a set of appropriate predator warnings figured in the ACSs of australopithecines.

Those who think predator calls are precursors of words may want to seize on this possibility and claim predation on human ancestors as the trigger for language. They should be cautious. Australopithecine alarms would have had exactly the status of alarm calls in the ACS of vervets. They would have been situation-bound, and their automatic linkage to stereotyped reactions—hiding in bushes, climbing trees, whatever—would have stopped anyone from using them in neutral, information-exchanging contexts. (Imagine yourself having to look for the nearest tree every time I said "saber-tooth"!)

However, alarm calls might have had one positive consequence.

When we looked at the ACSs of chimps and bonobos, we noticed the absence of any kind of call that wasn't either social or sexual. Even though alarm calls cannot be combined, and show no trace of either symbolism or displacement, they do at least direct the receiver's attention to some objective feature of the external world—unlike social or sexual calls, which merely express attitudes toward, or attempts to manipulate, other group members. And they are also arbitrary—there is no way the "leopard call," in and of itself, would evoke the image of a leopard or anything a leopard was likely to do. That's to say that though they don't behave like words, alarm calls do have two of the properties of words.

We saw in chapter 4 that trained apes, for all their undoubted intelligence, take a long time to "get it"—to grasp the idea that a signal could have objective reference—even though, once they "get it," learning new signals comes easily to them. That initial slowness may well be due in part to the absence from their ACSs of any signals that are even functionally referential. Even if alarm calls couldn't themselves morph into words, they might accustom their users to the notion that a signal could express more than mere feelings, needs, and desires. Such calls, in fact, could have helped get them ready for words.

THEM BONES, THEM BONES, THEM (NOT SO) DRY BONES

Meantime the millennia went on rolling by, and things were getting grassier and grassier.

Even while short-term climate changes caused things to yo-yo up and down quite a bit, the overall drying trend went on. Forests retreated eastward and upward, toward the summits of the Central African mountains that still caught moist air from Atlantic winds. Woodlands split and shrank, reduced to narrow galleries along shriveling watercourses, or vanished altogether. The grasses took over, rippling in the wind, lion-colored once the brief rains had passed. In this strange new landscape, our ancestors had to survive.

In all probability, they almost didn't.

For omnivores, savannas make living tough. Fruits and nuts become rare. The robust australopithecines, with their massive tuber-chompers, could still make a living of sorts. Apart from them, and apart from mammals and reptiles small enough to live on insects, all savanna dwellers must be either herbivores or carnivores.

Other primates that had adapted to savannas, baboons for example, had gone the herbivore route. But baboons had had millions of years in which to develop an appropriate digestive system. The gracile australopithecines didn't have the luxury of indefinite time. They needed a solution right there and then, or they'd go under.

Carnivory seemed like the only way to go. Since all great apes consume meat from time to time, this did not present a digestive problem. It did present logistical problems. In forests, for instance, chimps are able to kill monkeys by surrounding a tree with a monkey in it and cutting off all its escape routes. How do you do that in open grassland, where every degree in the 360 is a potential escape route? Superior speed, at least over short distances, is the answer for many australopithecine-sized predators. But whatever the advantages of bi-pedalism, four legs are usually faster than two.

Of course, there's always ambush hunting. A primate intelligence should have been good enough for that. And I'm sure australopithecines hung out around waterholes, hidden by long grass, motionless for hours if need be, waiting for the chance to pounce on a surprised animal before it had time to react. But when you've got that far, what do you do? If your prey's about rabbit size, you can wring its neck. Anything much bigger, you have problems. You don't have any of the right carni-vore equipment—strong jaws with sharp teeth to seize necks, hooked claws to rip out intestines. Try to kill even a small deer bare-handed, you'll see what I mean. And as far as we know, at this stage in our ances-tors' development they had no weapons whatsoever.

In fact I'd like to take anyone who still believes our remote ancestors lived by hunting, drop them in a modern savanna foodless and empty-handed, and see how long they survive. Well, not entirely empty-handed—I'm soft-hearted enough to leave them a cell phone so I could come get them when they cry uncle. It would make great TV, if noth-ing else.

Joking apart, there's just one kind of hunting they might have devel-oped, and that's endurance hunting.

We know that endurance hunting is still practiced by some of the few contemporary hunter-gatherers that are left. We don't know how far back it goes. An endurance hunter simply picks out an animal and runs after it. Of course, over a short distance the animal can run much faster. But it can't run indefinitely, and the hunter can.

That's one of the big payoffs of bipedalism—you may not be fast, but you're steady. It's unlikely that endurance hunting was what origi-nally selected for bipedalism—it must have taken hundreds of thou-sands if not millions of years to achieve the physical infrastructure, the stamina, and the muscular control to make endurance hunting success-ful. But once these things were in place, endurance hunting presented prey animals with a situation they may never have faced before. Most other predators burn out after their initial dash. If that dash isn't successful, they break off, relax, build energy for their next shot. Prey animals had no strategy to cope with a predator who, like an Energizer battery, just kept on going. Sooner or later they would collapse, semiconscious. Even if you had nothing to kill them with, you could just wait while they died of thirst and/or exhaustion.

The problem with endurance hunting is that while doing it you may end up as prey yourself. It's not something a large group can do, because with a single animal—and one small enough to run instead of just charging you—there won't be enough meat there to make it worth the group's while. So you alone, or at best you and a buddy or two, will be out there away from the protection of the group, maybe many miles away, maybe for days at a time, with no good means of defending yourself. Since you'll be in almost constant motion, you will quickly register on the gaze of any other predator, whose eyes will be constantly sweeping the savanna for its next dinner.

So while, in principle, australopithecines might have engaged in endurance hunting, I think it's doubtful anyone did until they had some kind of weapon, a spear maybe, that would put them on more equal terms with other carnivores. So far as we know, effective spears go back maybe half a million years, and we're talking now about two and a half million. A pointed stick, maybe? Come on, this is not the Monty Python self-defense-class skit. This is real life, late Pliocene style.

That left only one alternative: scavenging.

But there were problems with scavenging too.

We tend to think in terms of two classes of carnivore, hunters and scavengers—lions hunt, hyenas scavenge. Wrong: hyenas will pack-hunt, while most of the big cats scavenge when they get the chance. Predators lack any sense of sportsmanship—if they can eat without having to work for it, they'll do just that. Hunting is energy-expensive; it's what you do if there's no reasonably fresh meat already lying around.

So there's a natural hierarchy of scavengers. The big cats come at the top, naturally, above hyenas, wild dogs, and the like. But if there are only one or two big scavengers and plenty of smaller ones, the tables can be turned. Below them come the vultures, who will take care of pretty well anything the four-footed scavengers leave. Where in this hierarchy could the australopithecines enter?

Where do you enter any new enterprise? At the bottom, of course.

But what was left after the vultures had done their stuff? Well, hardly anything but the bones.

If ever a species needed to open up a new niche, *Australopithecus garhi* was it. And just such a niche was right there waiting for it—inside the bones, you might say. For inside the bones, inaccessible to any species without tools, was one of the richest and most nutritious foods known: bone marrow.

Smaller, more fragile bones can be, and often are, cracked and crushed by scavengers' teeth. Big ones are too thick, too strong. But a nimble primate with a hammerstone in his hand can break into the biggest bones. What australopithecine Einstein first figured this out, we'll never know. But sure enough, in recent years, cut marks made by primitive tools (and even a few of the tools themselves) have been found on bones at sites associated with *garhi* but too early for those tools to have been made by *garhi*'s successor, *Homo habilis*.

This has been seen as an embarrassment by some paleontologists, for the fact that *Homo habilis*—"handy man"—made tools, and the belief that no one else had done this, formed the main basis for making *habilis* the first identified member of the human family. And to be sure, the tools of *habilis*—the so-called Oldowan industry—may have been more sophisticated than the *garhi* tools. Or so the experts tell us. You or I, if we picked up one of either by chance, probably could not tell them from stones that had been cracked and shaped by natural forces—that's how primitive both lots were.

What matters, though, is that *garhi* and *habilis* faced the same challenge and (regardless of whether or not one was the ancestor of the other) dealt with it in the same way. And, looking at it from a general primate perspective, you could say this was no big deal. Chimpanzees on the Ivory Coast use unmodified (but carefully selected) stones to break open palm nuts. Homology or analogy? Who knows? Maybe the last common ancestor broke stuff open, or maybe it's just an idea that occurs spontaneously to any animal with a big enough brain when confronted by something hard with edible stuff inside.

But from our ancestors' perspective, breaking into bones had at least four big advantages going for it:

- Abundance: there were lots of herbivores on the savanna, so there were always plenty of bones around.

- Permanence: bones were not about to go away, like live prey; they would remain accessible for long after the demise of their owner.
- Lack of competition: no other animal could utilize this particular food source, so other scavengers would be long gone when our ancestors arrived on the scene.
- High-value product: nothing on the savanna was more nutritious, ounce for ounce, than bone marrow.

So first *garhi* and then *habilis* became low-end scavengers. And, lo and behold, brains began to grow.

Primate brains are bigger, relative to overall body size, than those of other mammals, and a necessary condition for that is a rich diet. (A sufficient condition is, once you've got that bigger brain, you have to find something for it to do, if it's to earn its high-energy upkeep.) Brains pretty much stabilized throughout australopithecine time, because an omnivorous woodland diet could barely keep level with a fruit-enriched forest diet. Bone marrow set in motion a trend that didn't reverse until quite recently, a progressive tripling of our ancestors' brains.

But that wasn't what started language. A bigger brain would indeed have come in handy once language had gotten started, and language itself would have selected for bigger brains. But for all the claims that "bigger brains made us more intelligent and that's how we got language," I've never seen even one backed by any kind of explanation of exactly *how* these developments came about.

For language, what you needed wasn't brains, wasn't even intelligence. Just the right kind of niche.

Meat, glorious meat!

Though bone marrow may have been rich, it wasn't really plentiful enough. Bones themselves may have been numerous, but the actual quantity of marrow each bone contained was quite small. However, there was another food source on the savanna that, while it might not carry marrow's nutritional punch, sometimes became available in quantities that stagger the mind.

That source was dead megafauna.

Let's look at the how and why of it. First of all, why were big animals there at all? The answer is the size niche. At the top of every order of being, there's a size niche, whether we're talking of trees (sequoias), ocean dwellers (blue whales), dinosaurs (sauropods), or mammals (mammoths). The size niche exists, permanently, within any order, simply because if you're bigger than anything else around, you're virtually invulnerable to attack. Nothing can become indefinitely large—constraints inherent in body plans, gravity, finiteness of food supply, and doubtless other factors prevent this. But in any order some animals will get to be as big as they can; natural selection guarantees it.

I've mentioned the large number of predators that roamed the savannas. Until a species arrived that could make weapons, size was the only real protection against these ferocious meat-seekers. So, in the savannas of two million years ago, efficient and widespread predation by carnivores selected for greater increase in size among herbivores. Indeed, there were several varieties of large herbivore: mammoths, deinotheriums and other predecessors of modern elephants, and the ancestors of rhinoceros and hippopotami. To their size they added a further line of defense: thick, leathery hides. These animals were perhaps the only ones that were often allowed the luxury of a natural death. And, even when dead, they enjoyed another day or two of invulnerability. For their skin was so thick and tough that, while a predator's teeth might be able to pierce it, they could not slice it or tear it open to extract the masses of meat that lay beneath.

Scavengers had to wait—pacing to and fro impatiently, or, more wisely, just lying in the long grass conserving their energy—until the action of bacteria inside the dead flesh released gases, and the gases expanded until they ruptured the dead animal's hide. Then, and only then, could the scavengers, in their order of precedence, move in for the feast.

That left an open niche, a narrow window of opportunity for any species that could cut the hide and access the meat before natural decay made it available to everyone.

Could our ancestors have opened up that niche?

Nicholas Toth, codirector of the Stone Age Institute in Bloomington, Indiana, and one of the leading authorities on prehistoric tools, set out to answer that question. With his wife, Kathy Schick, also a codirector of the Institute, and their associate Ray Dezzani, he took flint and lava flakes identical with those produced in the course of Oldowan toolmaking and they set about butchering an elephant that had died of natural causes.

It was a daunting task. "Initially, the sight of a twelve-thousand-pound animal carcass the size of a Winnebago can be quite intimidating—where do you start?" Schick and Toth begin their account. To move it would have required heavy-duty machinery—"You have to play the carcass as it lies." Schick and Dezzani started cutting anyway and were "amazed . . . as a small lava flake sliced through the steel gray skin, about one inch thick, exposing enormous quantities of rich, red elephant meat." And since "modern scavengers normally do not eat a dead elephant until it has decomposed for several days, such carcasses may have provided occasional bonanzas for Early Stone Age hominids."

Sure, but why only "occasional"?

The reason usually given is that very few megafauna remains have been found in what are known as "catchment sites." To understand what this means, we have to understand the difference between "catchment scavenging" and "territory scavenging."

Before two million years ago, most prehuman scavenging was catchment scavenging. Findings of artifacts and fossil bones, both prehuman and animal, cluster around particular locations—confluences of streams, rocky outcrops—making it seem that our ancestors used these locations as temporary or even semipermanent bases and scavenged in the area that immediately surrounded them. About two million years ago, a new strategy took over. Prehumans now ranged over broad territories, and instead of taking meat to a catchment site for processing, butchered and consumed it at or near wherever they found it. We'll see in a moment evidence that a correlated but much more significant change also took place around two million years ago.

So the fact that few remains of large animals are found at catchment sites tells us nothing about how often such animals were found and processed *after* the date when such sites fell out of use. Since dead megafauna could have cropped up anywhere, we would have to dig up the whole of East Africa to find out. Obviously, that's out of the question. We can only estimate how often our ancestors might have scavenged dead megafauna by getting help from modern statistics on large-animal populations.

Right now, the African elephant is an endangered species. However, there are still around a half million of them. These occupy a range of just over two million square kilometers, which means that, on average, there's an elephant for every four square kilometers. Before humans started slaughtering them for their tusks, we may reasonably suppose a density closer to one elephant per square kilometer. Or, in an area of, say, 150 square kilometers, 150 elephants.

An area of 150 square kilometers is seven and a half miles long and seven and a half miles wide. To a group of human ancestors at the center of such an area, a large part of it would have been visible to the naked eye, and any part of it could have been reached on foot in a couple or three hours, without hurrying. So it doesn't seem unreasonable to suppose that whenever a large animal died within that area, someone in the group would have spotted it pretty quickly.

Modern elephants in the wild live on average from sixty to seventy years. So within our 150 square kilometers, at least two elephants would have died every year. And that's just elephants. We haven't considered the ancestors of hippopotami, rhinoceroses, and any other large beasts that might happen to have been around. "Occasional bonanzas" would thus have occurred at least every couple of months or so.

All very well, you say, but deaths would hardly have been distributed evenly across an entire landscape. Maybe they all went and died near waterholes. And in any case, elephants don't stay still; they roam all over the place. For all your statistics, years could roll by without a single elephant dying in any particular 150-square-kilometer range.

That's very true. But it wouldn't matter, once territory scavenging had taken the place of catchment scavenging. Suppose hominids took to following megafauna herds, just as human hunter-gatherers in high latitudes would later follow the seasonal migrations of caribou or reindeer. Or suppose a larger group split into smaller groups, vastly extending the scavenging range, a range that could have been extended still farther as scavengers learned to read signs—dung piles and beaten trails, or better still, the circling of distant vultures. Then, far from "occasional bonanzas," dead megafauna could have provided our ancestors with a very substantial portion of their diet.

However, "could have" is a long way from "did." Is there any evidence our ancestors did in fact develop and exploit this new and quite unique niche?

Cut marks and optimality

The answer is yes. There are two quite separate but mutually reinforcing lines of evidence to indicate that they did.

The first comes from sequences of cut marks on fossil bones, the second from something called optimal foraging theory. Let's look at each in turn.

When you butcher a carcass with a sharp piece of flint or lava (or anything else for that matter), you inevitably leave cut marks on the animal's bones. This is true even if you're not trying to sever them. Bones just get in your way, as everyone who's ever carved a turkey knows.

In the same way, when a predator chews up a carcass, every now and then its teeth catch on a bone. The big cats of those days had sharp teeth, and those teeth made indentations in bones that are quite different from the cut marks stone tools leave.

Sometimes both animals and human ancestors worked (at different times, presumably) on the same carcass. You can tell when that has happened because one set of marks is superimposed over another. This shows which accessed the carcass first—prehuman or nonhuman.

Up until around two million years ago, wherever such pairs of markings are found, the cut marks of tools are always uppermost. In other words, other carnivores were getting at the carcass before our ancestors had a chance at it. Those ancestors were still entry-level—at the bottom of the scavenging pyramid, breaking bones for the marrow they contained.

Around the two-million-year mark, things change. Now, with increasing frequency, the sequence of marking on bones is reversed. Now it's the stone-tool cut marks that lie underneath, with animal bites superimposed on them. Our ancestors have somehow managed to move to the top of the scavenging pyramid. They're getting at the meat before anyone else has a chance at it. And the most likely, perhaps the only, way they could have done this is by accessing megafauna carcasses before anything else had a chance at them—in other words, by cutting through intact hides just like Toth, Schick, and Dezzani did.

Notice the period when the cut-mark sequence changes—it's around the time that catchment scavenging was replaced by territory scavenging. Could one change be the consequence of the other? It looks as though both were mere aspects of a larger process—the construction by our ancestors of the high-end scavenging niche.

Remember that two-million-year boundary and what lies on either side of it. Before it, there was catchment scavenging, and catchment sites with few if any bones of large animals. Naturally—just imagine slinging a mammoth leg over your shoulder and trotting off with it to home base. And in any case, if your main target was bones, you could afford to limit yourself to a relatively small home range, because bones were plentiful and didn't wander around.

But then prehumans moved into the high-end scavenging niche. Here, conditions were quite different. Your main targets, whenever you could get them, were megafauna carcasses. These could be lying around anywhere, and you had to go wherever they were: territorial scavenging, in other words. The farther you roamed, the more carcasses you might find. And they'd be too far, in most cases, for you to lug the remains to any kind of refuge. You'd have to sit down and consume them on or relatively near the spot.

Well, you may ask, why go to the trouble? Catchment scavenging had gone on successfully for hundreds of thousands of years. Why change now, even if you had learned how to cut through mammoth hides?

The answer is optimal foraging theory.

Optimal foraging theory was originally developed by the late Robert MacArthur (then at Princeton) and Eric Pianka of the University of Texas, Austin. (In a recent but unrelated development, creationists reported Pianka to the Department of Homeland Security for allegedly claiming that 90 percent of humans should be eliminated—Pianka claims he was merely warning that in a currently overcrowded planet, a mutated virus could do just that.) The theory states that any species will choose, out of available foods, just those that yield the highest calorific intake relative to the energy that's expended in obtaining them. Since MacArthur and Pianka wrote the first paper on optimal foraging theory more than four decades ago, countless studies of species ranging from gulls to stream insects to white-tailed deer have, with relatively few and usually explicable exceptions, supported this theory.

For modern humans, a supersized McMeal clearly represents the most calories for the least effort (evolution never having even imagined a species some of whose members would have virtually limitless access to food). For our ancestors, however, meat obtained from dead megafauna filled the bill. It was not as nutritiously rich as bone marrow, but unlike bone marrow it was available in vast quantities—in the carcasses of Winnebago-sized monsters on whose flesh you could feast for days on end. And you didn't have to work for it or hunt it. You just had to keep your eyes open and locate it, so energy expenditure would have been low compared with the caloric yield. For even if you had to look for a long time, there were always bones, plus the odd rodent, the odd tuber, the odd bees' nest to keep you going.

The only problem was the risk factor.

It wasn't only human ancestors for whom scavenged megafauna meat made the best nutritional bargain. Optimal foraging theory predicted the same outcome for any savanna-dwelling carnivore—for the big cats, the hyenas, the vultures, each and every one of the other scavengers. And were these animals, old hands at the game, going to let some Johnny-come-lately primate get away with what for millions of years they'd regarded as rightfully theirs?

They would kill the interlopers if they got half a chance.

How could our ancestors have coped with the competition? They had no natural defenses against their competitors' teeth and claws. They had no artifacts that by any stretch of the imagination you could call weapons. All they had—and that only potentially—was numbers.

Nathan Bedford Forrest was the least educated but the most innovative of Civil War commanders; for instance, he was the first to realize that the best use for cavalry was not to charge around waving sabers but to get men with rifles into commanding positions as quickly as possible. His was the pithiest-ever advice for success in combat, whether against human enemies or carnivorous scavengers:

"Git thar fust with the most men!"

If only our ancestors could come up with sufficient numbers, they could hold off the competition by screaming and flinging stones while they butchered and consumed the carcass. But how would they gather those numbers? A general like Forrest could issue orders and ensure that his orders were obeyed. But how, without language, could you get the most to get there at all? And was there, in the whole course of evolution, any kind of precedent for that?

We seem to have gone a long way from language, I hear you muttering.

Don't worry. In the next chapter, we'll go further still.

7

GO TO THE ANT, THOU SLUGGARD

Go to the ant, thou sluggard; consider her ways, and be wise.

—Proverbs 6:6

BEYOND VERTEBRATES

Several years ago, the prestigious journal *Science*, which does not normally pay much attention to language, published an article coauthored by Marc Hauser, Noam Chomsky, and Tecumseh Fitch entitled "The Faculty of Language: What Is It, Who Has It, and How Did It Evolve?" The article was placed in that section of the journal named "Science's Compass," and it was indeed designed to give directions to us poor benighted folks who (unlike the authors of the article) had actually been laboring in the quagmire of language evolution studies for a number of years. In chapter 9 we'll examine this article and I'll show you why, far from being a road map for future research, it is in fact a pernicious piece of misdirection.

However, the article did contain one useful piece of advice: "Current thinking in neuroscience, molecular biology, and developmental biology indicates that many aspects of neural and developmental function are highly conserved, encouraging the extension of the comparative method to all vertebrates (*and perhaps beyond*)" (my italics).

Well, let me caveat that, as Alexander Haig might have said. The part that follows the last comma is what's useful. The part that precedes it, what in the authors' opinion makes it worthwhile for us to look for language origins beyond vertebrates, relates to what is known nowadays as "evo-devo," the marriage of evolutionary and developmental biology. Evo-devo looks at the genes that turn a fertilized cell into a wasp or a mouse or a human and asks what our new and ever-growing understanding of genetic processes can tell us about how, out of such limited materials, so many varied life-forms evolved. One key insight of evo-devo is that homology is much more widespread than we thought.

In chapter 4 we looked at the difference between homology and analogy. Homology, you'll recall, was when the same feature occurs in two species because it was shared by their common ancestor. Before evo-devo, people looked for homologies only among closely related species. It wouldn't have occurred to anyone, for example, to choose the wings of birds and bats as examples of homology. To find the common ancestor of birds and bats you'd have to go back around 300 million years, and in both lines of descent you'd find innumerable intervening species that didn't have wings. So wings just had to be analogies, the result of aerodynamic factors—how else would an animal get to fly around? And the genes that set up the two different pieces of equipment for flying just had to be different from one another—didn't they?

If you got that one wrong, you're in good company. Ernst Mayr, doyen of twentieth-century evolutionary biology, wrote in 1963: "Much that has been learned about gene physiology makes it evident that the search for homologous genes is quite futile except in very close relatives. If there is only one efficient solution for a certain functional demand, very different gene complexes will come up with the same solution, no matter how different the pathway by which it is achieved. The saying 'Many roads lead to Rome' is as true in evolution as in daily affairs."

Since then we've learned an awful lot about genes and how they work. In particular we know that evolution, like any good home-workshop do-it-yourselfer, never throws stuff away. No matter how useless some bit of machinery looks, there's no knowing when a use for it might come up. With a little ingenuity, you can reshape it and use it all over again—saves you going out and buying something. And in nature there's no Home Depot to go to anyway.

Thus many of the same genes are still there in bats and birds and those same genes regulate the development of the body parts that still underlie their superficially different wings—the arm bones and the bones of hand and finger that those same genes also developed in all the forelimbs of all the ancestors of bats and birds that didn't happen to fly. That's what's called in the trade "deep homology." In fact, deep homologies can go even deeper than that. To find a common ancestor for the mouse and the fly, you'd have to go back almost twice as far as the common ancestor of birds and bats. And yet the same set of genes determines the overall body plans of mice and flies.

For there's no simple one-to-one relationship between genes and body parts, in which the same genes stamp out identical, cookie-cutter, one-size-fits-all bits and pieces; genes can be differently expressed in different combinations and in different environments, giving rise to body plans that can look strikingly unalike until you see that the same basic proportions hold in each of them—back to front, side to side, inside to outside.

Giraffes' necks are many times longer than ours, but they have just the same number of bones—seven. Genes that govern bone length make bones longer or shorter, to fit the overall blueprint for whichever animal they're building. But they can't change the number of bones—that's not their job, and it seems it isn't any gene's job, unless some strange mutation just happens to come along.

There's a big problem, though, in relating any of this to language (as I believe Noam Chomsky wants to do, in ways that are, as yet, far from apparent).

You see, deep homology doesn't shape behavior. It helps to shape body parts, no question of that, but I don't think it's ever been claimed that a particular *behavior* results from deep homology. And language is, after all, a form of behavior—behavior underpinned by genes, to be sure, but in no way controlled, determined, or mandated by them. Genes help to shape body parts (including brain parts, naturally), and body parts help to shape behavior, but there are just too many independent variables between genes and behavior for deep homology to help when it comes to looking for antecedents of particular behaviors— especially of a behavior that's unique.

So, for all the promise of evo-devo in other areas, it gives us no good reason to go looking for precursors of language in strange new places. If we're to search outside the primates, even outside the vertebrates, it's behavior, not genes, that we'll be looking for. Because that's what niche construction theory is telling us to do.

New behaviors arise through the construction of niches, and it's those niches that in turn determine how genes will express themselves. (It was, after all, the development of powered flight that eventually caused genes normally devoted to building front legs to express themselves, among birds and bats, in the form of wings.) Behaviors develop because the niche requires them and can't be constructed without them. Species have a choice: stick to the old ways and maybe go extinct, or try branching out into something new. They may not succeed in this second course. They may simply lack a genome that can be tweaked into supporting the behaviors they'll need—they're variation-limited, in Eörs Szathmáry's phrase. But if they're only selection-limited, they'll respond to the selective factors inherent in the new niche by constructing new and niche-appropriate behaviors, and the feedback process between genes and behavior will kick in.

Accordingly, we're not looking for deep homologies, we're looking for niche analogies: niches that share the same kinds of selective pressure. Since it's the niche that matters, the niche that determines the animal's behavior, it doesn't matter how far we stray from our own species. Whatever kind of living organism we're dealing with—reptile, mammal, fish, insect, bird—the same kind of niche will have similar consequences.

Remember at the end of chapter 2 I suggested a selective pressure highly likely to lead toward language: the need to transmit information about food sources that lay beyond the sensory range of message recipients. So what we need to look for, in the vast array of species on earth, are ones that have niches requiring this kind of information exchange. If the information that's transferred happens to concern food sources too large to be handled by individuals, calling for some kind of recruitment strategy, so much the better.

Surprisingly enough, almost the only species that meet these criteria are ants and bees.

AMONG THE HYMENOPTERA

People have known about bee ACSs for a long time, ever since Karl von Frisch's studies of honeybee communication half a century ago. It was known, even then, that the ACS of bees was capable of displacement.

Recent research has more than confirmed von Frisch's work, adding fascinating knowledge about how bees measure distance (they compare
the speeds at which images of objects in the landscape appear to cross
their retinas as they fly).

But beyond the "Wow!" factor, people didn't think bee displacement had anything to do with language. Bees were too far from us, phylogenetically speaking. And since behavior, like anything else, was seen
as a monopoly of the genes, there couldn't possibly be a connection. It
wasn't until niche construction theory came along that you could look at
the situation in any other way. But when you did, the picture changed
dramatically.

So, let's look at bees and how they behave and why.

Bees are eusocial; in every community, only one female, the queen,
and a handful of males, the drones, are fertile and can mate. This means
that all other members of the community are sterile and are also siblings
of one another. Since in addition they are haploid-diploid (the female
parent has two copies of each chromosome in each cell, the male only
one) bees share far more of their genes than most siblings do. Now recall
the principle of inclusive fitness, meaning that animals will try to preserve not only their own genes but those same genes wherever they
occur. Remember how, in the case of survival calls, animals risk their
own lives to benefit close relatives who share their genes. So, just as
you'd expect, there's more cooperation and cohesion among bees than
in almost any other species, until you get to humans.

If the good of one is the good of all—and it is, for if they don't store
enough honey all the bees will die next winter—then it's in the interests
of everyone to share information about food sources. This is not so
among many species. Members of most, on finding a tasty food source,
keep it to themselves, and a few may even use phony warning calls to
distract group members who try to take it. For bees, food sources are
patches of flowering plants that bloom all at once and may not last more
than a day or two. If a single bee locates such a source, it can't hope to
fully exploit that source. For the good of itself and all, it must recruit its
nestmates to help it.

Recruitment—that turns out to be the key word in the birth of language.

The sites for which bees must recruit nestmates may lie several kilometers from the hive. A measurable period of time, several minutes at least, must elapse between when the bee locates the source and when it passes on the information. Therefore an effective bee ACS must displace—it must transmit information about states and events existing in a different place and at a different time. Unlike other ACSs, it cannot function if it remains imprisoned in the here and now. But in escaping the here and now, it is responding to the selective pressure noted at the end of the last section—the pressure likeliest to move in the direction of language.

In order to recruit effectively, bees need to tell other bees where food sources are located and how far away those sources are. They show whether the food site is near or far by choosing between two kinds of dance. If the site is within say seventy-five yards of the hive, they dance in circles, so that's called the "round dance." If it's farther, they dance in a series of elliptical loops, waggling their bodies as they come through the straightaway in the middle, so that's called the "waggle dance." The faster they dance, the farther away the site is. The axis on which they waggle does not, as you might suppose, point toward the food; dances take place on the vertical face of a honeycomb, and even if they were done on the level, by the time a bee got out of the hive it would have no idea which direction to go in. (Imagine yourself being pointed in a certain direction in a dark room in a windowless office block and then, after making your way through several halls and corridors and down a couple of staircases, trying to recalibrate that orientation in the daylight outside.)

So the bees execute an amazing transformation based on the current position of the sun. They compute the angle between the sun's current position and the site and convert that from horizontal to vertical, representing the sun's position by a vertical axis and the angle between sun and site by the angle between the vertical and the axis of the waggle dance. If this sounds easy (and I don't think it does, do you?) go to a south-facing window, pick an object, estimate its angle from the sun, and then take a pencil (*not* indelible!) and mark that same angle as a variance from the vertical on the nearest wall. (No mechanical aids, not even your fingers—that's cheating! Bees do it all in their heads.) Then think: a bee's brain is pinhead-sized compared to yours.

But I *thought*, you complain. The bee didn't. It just used instinct.

There you go with that homocentric prolearning bias again. Now tell me how you produced the sentence you just spoke. That was an unconscious mental computation, and so was the bee's reading of her buddy's waggle dance. Both are just cases of subconscious thought.

Did the first-ever bee get born with that instinct? The first bee wasn't even social; it was every bee for itself, back then. At some time in the past, bees somewhere must have constructed the eusocial niche, and must later on have started, in some doubtless highly clumsy and inefficient way, the practice of informing their fellows where to go for food. So they got it wrong most of the time? Of course, but with just a few successes some bee colonies survived the winter while others went under. The survivors got better at it as the feedback mechanisms of niche construction kicked in, the niche shaping how the genes were expressed (and selecting rare beneficial mutations as these came along) and the expression of the genes shaping the niche, until, probably millions of years ago, you reached the spot-on computations of modern bees.

So honeybees are the obvious model for a system of communication that involves displacement. But it doesn't follow that they're the best model. True, they're extractive foragers, just like our ancestors were. But bees forage in the air; our ancestors foraged on land. Bees forage in one place only, inside flowers, for two things only, the nectar and pollen those flowers contain; our ancestors foraged in many different kinds of places, for many different kinds of food. They may have preferred the meat of large mammals, but that was a chancy and unpredictable business. Between bonanzas they had to go the omnivorous route of *their* ancestors, the australopithecines.

In their foraging, our ancestors were less like bees than like ants.

Ants 'r' us?

Ants have always fascinated us. Traditionally, they are held up as models of frugality and industry. Aesop's fable contrasts the responsible ant with the irresponsible grasshopper, fiddling all summer, starving when winter comes. The epigraph to this chapter is but one of a number of references to ants in the Bible, the Koran, and other religious texts. And at the end of this book we'll note some eerie parallels between ants and humans, suggesting one far-from-inconceivable human future that most of us would rather not conceive.

For now, however, their communication is our sole focus.

Like bees, ants are extractive foragers; working from a central place, the nest, they radiate out over relatively large areas. Their diet varies from species to species, but many species will eat almost any organic matter that doesn't move, and quite a bit that does, if it's not too big and lively for them. They will eat other insects, other ants, dead caterpillars, overripe fruit, you name it. They're sweetaholics, as you know if you've ever spilled sugar in the kitchen, but they love protein too. A lot of the things they love are much bigger than they are.

Ants, again just like bees, are eusocial. Beyond that, since there are more than eleven thousand known species of them, it's hard to generalize. Some save food for the winter, just like Aesop's did, and some don't. Some make war on the nests of other species; most don't. Some herd, some farm, some forage, just like humans over the last few thousand years. But all of them communicate chemically.

Ants produce, from their tiny bodies, a stunning range of chemical substances, some of which our own chemists have not yet replicated. For instance, one species of ant that conquers other colonies and enslaves their inhabitants (sound familiar?) can release a chemical that causes ants of the colony they're attacking to fight one another. (I've dreamed of creating a human equivalent, to be produced under the trademark "Fight!" What's deterred me—apart from ignorance of neurochemistry—is the fear that, bearing in mind the IQs of military leaders, they might use it on our own troops: "Fight? We don't want the enemy to fight! We want them *not* to fight! We want *our* troops to fight *harder!*") There are chemicals that distinguish nestmates from interlopers, that attract mates (for the few that can make use of them), that warn of danger, that cause ants to congregate; in higher concentrations, the last chemical indicates that a nestmate has been injured and throws the ants into attack mode, ready to sacrifice their lives to protect the colony. But the chemicals that concern us here are those used for communication, and the question to be answered is not "What?" but "How?"

To the extent that ants function as predators and scavengers (rather than fungus farmers or aphid herdsmen), their food sources are transient, unpredictable, and scattered. In order to feed all the inhabitants of the nest, the only workable method is a fission-fusion strategy.

And that's one thing that all our species of interest share. We've seen how chimpanzee groups will split and reform, although in the course of a day's foraging they will seldom be out of visual or aural range of one another. For them, it's more a case of, well, since there's not enough fruit on this tree for everyone, let's you and me move on to the next one.

Our ancestors had much stronger reasons for a fission-fusion strategy, since even back in australopithecine days food sources were almost certainly more meager, more widely scattered, and more ephemeral than those in the forests of the apes. Even if they gathered at a common site to sleep, there must have been many times when group members were out of sight and earshot of one another, maybe even miles apart.

Similarly, bees and ants forage either individually (unlikely in our ancestors, for security reasons) or in small groups. Moreover, and this is something that turns out to be quite rare among animal species—it's certainly not found in any primate other than us—their prey is often much larger than they are. Like Lilliputians with Gulliver, some ant species can overrun and capture small birds, lizards, and the like by sheer weight of numbers. But those numbers have to be recruited—"If you don't recruit, you get no loot," as Johnnie Cochrane might have put it.

Even where the prey is already dead or inanimate, like a fallen fruit, it is subject to decay and to scavenging from other species (although ants seldom have to face competition from larger and fiercer scavengers in the way human ancestors had to), so it has to be exploited as soon as found. Under these conditions, a recruitment strategy is inescapable.

Really? Couldn't ants and bees survive without it? It's possible that they could. But what's absolutely certain is that, if there were bee or ant colonies that practiced recruitment and bee or ant colonies that didn't, those that practiced it would prosper spectacularly at the expense of the others, and whatever genes supported recruitment strategies would spread throughout the population. In consequence, it's no surprise that such strategies have been adopted by probably all bee and ant species except for the relatively few that don't live socially.

Among the ant strategies are a couple of things that look surprisingly similar to two of the major building blocks of language: concatenation and predication.

We saw in chapter 2 that one thing an ACS couldn't do was to concatenate: that is, put two units together and thereby make something different in meaning from what those units meant when produced separately. But there's a species of ant, *Camponotus socius*, that has a complex recruiting behavior. Suppose one of these ants makes a food discovery. It returns to the nest (unless it first meets other ants), meanwhile laying a chemical trail so it can find its own way back. On meeting nestmates, the lead ant draws their attention by a characteristic shaking of its body, signifying a substantial food find, then turns and runs back along the trail, with the others following. As it runs, it continues to pump out the trailblazing chemical.

It occurred to Bert Hölldobler, a professor of biology at Arizona State and one of the leading researchers into ant behavior, to see what would happen if he removed the lead ant after its shaking display but before it had taken its recruits anywhere near the food. Nothing happened. The ants quickly lost interest and wandered off in different directions.

Next question: Why did they stop following the original trail? Because the leader wasn't there to follow, or because a fresh chemical trace was no longer being laid down? In some other ant species, the leader's physical presence is crucial. That's in those species that practice a form of recruitment called "tandem running": one ant grabs hold of another and literally drags it along the trail toward the food source. So Hölldobler manufactured the trail tracer from a mix of ingredients in the ant's bladder and its poison glands, removed his leader again, and laid down the trace himself ahead of the ants. Sure enough, they followed it.

Clearly the followers needed the shaking display followed by a continuous chemical flow from the leader to conclude that dinner lay somewhere down the road. The original light trail laid by the lead ant for its own use wasn't enough, even after the shaking. This isn't quite like joining words. The shaking and the chemical trail may be meaningless by themselves, so it's more like the joining together of in-themselves-meaningless speech sounds that we do to make words. But it's still concatenation of a kind; a kind that, primitive as it is, is found seldom if at all among the behaviors of other species.

As for predication, you'll recall that it's the basic linguistic act of taking some entity and saying something about it: "Dogs bark," "Birds fly,"

"Water's wet." Several species of the genus *Leptothorax* have developed something that, while far from the kind of predication we find in language, is nearer to it than anything in any other species. Here's one account:

> When a forager finds a plentiful food source it returns to the nest and regurgitates food to its nest-mates. Then it raises its gaster [the rear section of its stomach] and extrudes its sting bearing a droplet of liquid. This attracts nest-mates to it in the nest. As soon as the first of these nest-mates reaches the caller, the caller runs out of the nest and leads the nest-mate to the food source.

Note that this particular recruitment strategy involves the act of regurgitation. Many species regurgitate to feed their young, but that can hardly be what's happening here. Rather, it's as if the ant is presenting a sample of what's available—"Hey guys, here's what you'll get if you follow me." As we'll see in the next chapter, this strategy, explaining what kind of food is available, may be crucial for prehuman recruitment. Without knowing what they were being recruited for, it's doubtful whether protohumans would have allowed themselves to be recruited at all.

At the other extreme, bees don't need to sample the product; they already know what they'll be getting—pollen or nectar. Ants come somewhere between bees and humans in this respect: they consume a wide variety of foods, and they may be influenced by the type of food regurgitated to go out in greater or lesser force. I don't know if the experiments have been done yet, but they would be easy enough, using different foods, plus a control where the ant was somehow prevented from regurgitating before it raised its gaster. (If it refused to raise its gaster unless it had first regurgitated, that would be significant too.)

It may seem overly fanciful to regard the whole sequence—regurgitation plus chemical signaling plus tandem running—as the ant equivalent of "Come, such-and-such food is thataway." Moreover, the sequence, like the calls and other units of ACSs, is primarily designed to get creatures to do things (manipulation), and does so, just like ACS calls do, by means of a stereotyped set of behaviors that cannot be varied or altered in any meaningful ways.

But it must still be borne in mind that the recruitment sequences I have described are more complex than any communicative behavior in any other species (barring our own) and involve transfers of information more detailed and specific than any other ACS can perform. Moreover, the information transferred is not information about the here and now—as it is in the case of predator warning calls—but refers to things outside the sensory range of message recipients, just as (most of the time) language does.

RAVENS FLY IN, BUT, ALAS, DON'T TELL US MUCH

The question that arises at this point is, are ants and bees the only species that need to practice recruitment?

In the narrow sense in which we've defined it—which involves not just telling others you've found food, but bridging a time-and-space gap those others couldn't have bridged for themselves—recruitment turns out to be surprisingly rare in nature. It's limited by the fact that candidates must be

- Social: solitary species don't have to tell anyone anything.
- Foragers over an extensive range: most species travel relatively short distances to feed.
- Fission-fusion foragers: many species with extensive ranges still keep their usual social groups intact during foraging; consequently, whatever one knows, all know.
- Foragers for bulk items: the desired food source(s) must be too large or too well defended (or both) for individuals or small groups to handle.

The only species I've so far found that meet these criteria, apart from ants and bees, are ravens. (Thanks, Tecumseh Fitch, for turning me on to them.)

You can read fuller details about ravens and how and why they recruit in a fine book, *Ravens in Winter*, by Bernd Heinrich of the University of Vermont. Here is a very brief summary.

During winter, a major source of food for ravens comes from the carcasses of animals, many of whom have died of starvation. The locations of such carcasses are, of course, quite unpredictable (as are those of dead caterpillars, dead elephants, and the like). Mature ravens pair-bond, and if such a pair locates a carcass they will settle on it and stick around until it's consumed, driving off any other raven that tries to share it. (Note the competition factor, present in the prehuman megafauna-scavenging situation, although not significant for ants or bees.)

Immature ravens forage solo but gather in trees to roost at night (a typical fission-fusion strategy). If in its foraging an immature raven finds a carcass already preempted by a bonded pair, it doesn't stand a chance. And, at any given time, all the available carcasses in the area may be guarded by vigilant pairs.

However, if that lone searcher can recruit helpers, it can drive off the pair and access the carcass. But it can't do this without some way of telling its roostmates about the bonanza it has found.

Apparently it has a way. Normally, if no one has found anything, ravens don't follow one another when they go foraging. After a night of roosting together, they all fly off in different directions. But if one has located a carcass the previous day, several if not all of its fellow roosters will take off and follow it to its destination, fighting the current owners, driving them off, and sharing what's left of the carcass among themselves.

How do ravens do this? Nobody knows, yet. And it would be extremely hard to find out. You'd have to climb into the treetops (or better, let long-range infrared cameras do it for you) and then somehow determine, out of all the caws and pecks and wing flaps, which one (or ones) carried the crucial message. But even without knowing the mechanics of it, we can be pretty sure that ravens do have an ACS that has somehow achieved displacement.

It would surely be helpful to know how a species intermediate in its mental powers between ants and humans does it. And it would be helpful too to find more species that face the same problem as ants, bees, ravens, and human ancestors have had to confront. This is one of those times when that all-too-often-repeated saying "More studies are needed" isn't just an excuse for not making up our minds. Pending knowledge of other species, we'll just have to soldier on with the ants and the bees.

However, some of you may still be reluctant to accept that our ancestors could have started language by behaving like ants. For this reason, in the next section I'm going to act as my own devil's advocate, lay out all the arguments I can think of against the ant/bee scenario, and then show that all of them can be answered.

Devil's advocate

- The "languages" of ants and bees are evolutionary dead ends; tens of millions of years later they haven't developed into anything more ambitious, whereas language, something that can hardly have started more than two million years ago, is already a system of immense complexity and seemingly limitless productivity.

Well, what would you expect when one lot have brains smaller than pinheads while the other lot have brains as big as coconuts? Besides, any communication system will fulfill its owners' needs and no more than those needs. Ants don't need to gossip—they don't even have personal lives to gossip about. They don't need language for sexual display—most of them can't even have sex. They don't need it for Machiavellian strategies to enhance their own power and status—their power and status are irrevocably fixed at birth. So why would they develop their "language" any further?

- Bee or ant "languages" and human language are apples and oranges—the first are hardwired, the second culturally learned. A closed, hardwired system can't develop into an open, culturally mediated one.

No one's saying it could. The reverse is another matter. The production of the first human utterances can only have been spontaneous, not supported by any dedicated infrastructure. Nowadays, language is produced automatically, and speakers are totally unaware of the means by which it's produced—just as ants are unaware of the chemical mixes and fixed-action patterns that they use in their recruitment strategies. In the same way, the first attempts by ants to recruit helpers for foraging must have been spontaneous behaviors that only later were refined and perfected and absorbed into the nervous system by a process known as the Baldwin effect. Instinct is just fossilized behavior, regardless of whether we're talking of ants or humans. Changes in behavior trigger changes in genes at least as often (perhaps far oftener) than genetic changes trigger behavioral changes.

- Ant and bee "languages" are rigorously restricted to a single semantic area—the gathering of food. If human language began with the same function as ant language, why wouldn't it have remained as a narrowly restricted mechanism for improving foraging capacities, and never acquired any broader functions?

When I first encountered this objection, my first reaction was to point out differences in brain size and to claim that the first words were spontaneous inventions; once you had a brain big enough and flexible enough to name just one thing, you could name anything. Then when I reconsidered it, I thought, maybe they're at least partly right. Maybe for tens or hundreds of thousands of years after the first wordlike signals appeared, the modified ACS remained, just like an ant ACS, mired in the business—a vital one, you have to admit—for which it had been originally developed.

If this were so, it would resolve two other questions that language evolution studies have to face. One is why, if language started two million years ago, did it take so long to develop to its present degree of complexity? The answer to that would now be, because for an indefinitely long period it wasn't and perhaps couldn't have been used for anything but scavenging. How protolanguage might have escaped the scavenging trap and branched out into the world is discussed in chapter 11.

The other question is why, once some simple protolanguage had gotten started, did culture seemingly stagnate, almost up to the point where our very own species, *Homo sapiens sapiens*—wisest of the wise—emerged?

This to me is the more important of these two questions, as well as being one that most studies of human evolution either ignore or fudge. It's mind-boggling when you think about it. Shortly after the big-mammal-scavenging phase we reached at the end of the last chapter, our ancestors began to produce something called an Acheulean hand ax—a teardrop- or pear-shaped stone object that approaches perfect symmetry. Most paleoanthropologists think that Acheulean hand axes were used as tools, for chopping and/or cutting meat, but some think they were projectiles, others that they were simply the cores left after flake tools had been struck off them, and still others that they were an item of display made to capture the hearts of females. Whether it served any or all of these functions, this Lower Paleolithic Swiss Army knife was produced, virtually unchanged, for at least a million years.

When I give talks on evolution, I often tell my audience things like, "The new model Ford brought out this year is so good it will probably still be in use a million years from now." That helps to bring home to them the immensity of the gulf between our ancestors and ourselves. It's unthinkable that our species would produce the same model car even for a decade, let alone a period five orders of magnitude longer, no matter how good it was. Our itch for innovation—even if sometimes the new thing turns out worse than what went before—makes any such possibility ridiculous. Ancestors or not, the hand ax makers must have been a totally different kind of being from us.

The difference lies, of course, in the mind alone. In our physical being, in our emotions and our drives, I'm sure we're very close to them. There was never any moment at which you could take a parent and a child and say, "This child was a true human, but the parent wasn't." Yet somewhere along the way, our minds changed, and they changed quite quickly, as these things go in evolutionary time.

But if language, true language as we know it, was what reshaped the human mind, as I'll argue in chapter 10, there's a simple and straightforward explanation. Protolanguage did indeed stick, for a long time, at the bee/ant level, or only a little beyond it. What changed it, and how it changed, we'll look at in greater depth in chapters 11 and 12.

- How can you compare communication systems that use chemical signals, or shapes sketched in space, with a system that uses either sounds or manual signs?

Very easily. What such a system works with is quite immaterial. We saw in chapter 1 that ACSs use an immense variety of media—sounds, smells, gestures, lights—but that all these systems are saying the same kind of thing. What matters is not the medium but the message. Language sends the same messages regardless of the medium. You can use flags (did you ever see Monty Python's semaphore version of *Wuthering Heights*?) or the dots and dashes of Morse code: the same rules hold as in the spoken version. If bee or ant systems are sending messages that go beyond the usual ACS messages, that contain at least one feature that is otherwise found only in human language, then any difference in the means by which those messages are sent is quite irrelevant.

- When all's said and done, we're dealing here with a system of signals that form a closed (and very small) set of instinctive behaviors (meaning they can't be voluntarily added to) that are not real symbols, since they can be used only in a narrowly defined situational context (in contrast with words, which can be used in any situational context). How could things like these possibly be precursors of language?

They're not—at least, not in the sense in which some abilities of cotton-top tamarin monkeys are precursors of language. Tamarins can distinguish speech sounds from nonspeech sounds just as efficiently as newborn humans, though neither babies nor monkeys can do this when the same sounds are played backward. So here we have a (presumably homologous) precursor of adult human speech discrimination. But in this chapter we weren't looking for "precursors" in the sense of "things that might ultimately, sooner or later, have turned into language, or been used for language." What we were looking for were ACSs with features that breached at least one of the rigorous constraints binding on almost all ACSs. We were looking for abstract models of how language might have developed, not for precursors in the deep-homology sense.

Language isn't just unique—it's unnatural. The question is not so much why our species got it and no other did, it's why *any* species got it at all—why every species since the primal bacterium didn't go on happily using its ACS until the world got swallowed by an expanding sun or frozen by a dying one.

What a different world we'd be living in (or rather, wouldn't be living in) if that were so!

But it's not. For good or ill (it could turn out to be either), we somehow got language, and the only way I can see that we or any other species could have got it is by constructing some kind of niche that, by its very nature, forced us to break out of the prison of the here and now.

BACK TO THE PALEOLITHIC

It's time now to leave the hymenoptera to their fascinating yet strangely restricted lives. On the savannas, the forerunners of humanity are in motion. Let's get with them. It's a risky procedure—not for you, dear reader, but for me. To give you the full impact of what happened I'm going to have to go out on a limb, not too far, I hope, but a little beyond what the known facts will fully support. So bear in mind, certain aspects of what I'll describe may be proven, by subsequent research, not to have been completely accurate. But in essence, something like what I'll describe to you has to have happened. We know our forebears scavenged large dead animals, we know they faced fierce competition from other scavenger/predators, and it's a reasonable inference that they could have succeeded only by recruiting sufficient numbers. And, as I see it, there's only one way they could have done this—the same way ants and bees did it, by breaking the ACS barrier and achieving displacement.

So let's go back and see what could have happened, somewhere around two million years ago.

8

THE BIG BANG

What evolved into what?

"If apes evolved into humans, how come there are still apes?"

I was having breakfast, listening to a radio talk show; I nearly fell out of my chair. The caller was dead serious, though. Moreover, his tone had that gotcha smirk in it, like he'd just demonstrated irrefutably what dumb clucks all those highfalutin' professors were—they'd never thought of *that*, had they? I was so stunned I didn't hear how, if at all, the talk show host responded. But sure enough, that same week in the letters column of a giveaway newspaper the exact same argument popped up (fortunately in the next issue another letter-writer provided readers with the hundred-word version of Evolutionary Biology 101).

What, I wondered, is our educational system up to? Could the old Dover school board have made it any worse? And then I thought, Uh-oh—maybe everyone concerned could plead mitigating circumstances.

You see, our caller's flawed understanding of the ape-human transition—that the one was supposed to have just, uh, turned into the other—was a perfectly correct understanding of what most experts used to say, and a few still say, about the transition from one ancestral human species to the next. As far as the apes, the great tree of evolution branched every which way; then we came out of the branches, just as in real life we came out of the rain forest, and there we were, out on a limb, at the end of a long bare branch with nary a twig on it.

Australopithecines turned into *Homo habilis*, *habilis* turned into *erectus*, *erectus* turned into primitive *sapiens*, *sapiens* turned into Neanderthals, and Neanderthals turned into us. You could see this in the human family trees shown in old textbooks—gorillas and chimps and orangutans all obediently branching, and then a long straight line going right through four or five species to a dead stop, us. We might be descended from apes, but those diagrams at least put a decent distance between us and them.

Then, to the embarrassment of many, more and more quasi-human species began to show up in the fossil record. Clearly our branch *had* branched, after all, but for a while you could still save the day by dismissing these side branches as botched jobs, subhuman discards that at least had the decency to seek a quick extinction, while the unbroken human line marched triumphantly onward and upward.

As recently as the beginning of this century, a serious debate between the "Out of Africa" and multiregional hypotheses was still raging ("Is 'Out of Africa' out the window?" a headline in *Science* demanded). The "Out of Africa" hypothesis (based largely but by no means entirely on mitochondrial evidence) held that our species evolved in Africa between one and two hundred thousand years ago, subsequently replacing all other varieties of *Homo*. The multiregional hypothesis (based on claimed physical similarities between humans in specific regions of the world and prehumans in those same regions) held that prehuman species in all parts of the inhabited world, *erectus* and Neanderthals and primitive *sapiens* alike, had undergone an "improvement in grade," whatever that was, and that all, more or less simultaneously, had evolved into us.

You couldn't blame our caller for being confused. After all, the process he thought evolutionists upheld in general—one species turning into another—was the exact same process that evolutionists *had* upheld, and a few still uphold, when it came to our own species and its immediate ancestors. If the multiregionalists were right, and if what they believed went back to the dawn of humanity, then there *shouldn't* be any apes!

Nowadays a majority in all the relevant sciences accepts "Out of Africa." It's hard to see how anyone could do otherwise, given the known ways in which biological evolution works. The old one-straight-line-to-humans family trees are gone. But, amazingly enough, the conventional, branching ones didn't replace them for long.

As the human family multiplied, as *Homo heidelbergensis, Homo ergaster*, *Homo antecessor*, *Homo helmei*, *Homo rudolfensis*, and now *Homo floresiensis* appeared upon the scene, folk gave up on family trees. Now what you typically see are separate plain or colored blocks with names attached to them, scattered haphazardly over a six-million-year space, parallel to one another, with some overlapping and some not, sometimes with spidery lines connecting some of them but most of the time showing you nothing whatsoever about what evolved into what.

All these *Homo*s are supposed (at least by some) to be separate species. Bear in mind that not many paleontologists accept all of them as genuine species, very few would agree on the same list, and probably no two would totally agree on which fossils belong in which species. In part that's because an old saying among paleontologists (but it applies in any field of study) divides humans into two kinds: lumpers and splitters. If you're a lumper, you want to put several superficially different types into the same category; if you're a splitter, you want a separate category label for every type. Is there any way to get beyond lumping and splitting?

Yes there is, but you have to stop obsessing about bones and their minuscule variations and start looking at biological evolution, in particular at speciation and how it comes about.

I'm going to spend the next few pages talking about speciation.

Is this just a detour? Are we going away from language again? No way! Once again we must take our bearings from Dobzhansky's dictum: "Nothing in biology makes sense except in the light of evolution." And speciation lies at the heart of evolution, whether it be evolution of language or evolution of anything else—get that right and all the stones and bones obediently fall into place. Darwin didn't call his book *The Origin of Species* for nothing, even though the book is about "descent with modification" and not really about how any new species becomes distinct from other species. He knew that speciation was where the rubber meets the road, even if it would take more than a century for people to begin to get a handle on it.

The birth of language was just part—*had* to be just part—of what is often misleadingly referred to as a "speciation event." In evolution, most of the really interesting stuff happens when one species branches out from another and sets up, so to speak, in business for itself.

ONE MODEL OF SPECIATION

Crucial as speciation is, it's still far from completely understood; it's still possible for creationists and Intelligent Designers to use it as a wedge issue. Both creationists and IDers make a distinction between microevolution and macroevolution. Microevolution's fine by them, even natural selection, just so long as it stays micro. Things get wetter or drier, hotter or colder; naturally, already existing species adapt to those or any other environmental changes. What creationists and IDers balk at is macroevolution, the emergence of new species. There is where they see the Designer's fine hand, and the rarity of forms intermediate between one species and another doesn't help much.

So let's see what we do know about speciation, and then we'll see what it looks like when viewed from the perspective of niche construction.

People often use the expression "speciation event," as if you could buy tickets for one and still get home by dark. I've been guilty of using that term myself. But I was fortunate enough to be at a meeting where Robert Foley and Marta Lahr of Cambridge University gave a talk on speciation that for some reason was not included in the published proceedings. A pity; it was an eye-opener for me, and deserves a far wider airing.

According to Foley and Lahr, speciation, far from being an event, is a process that may span as long as a million years or more. (Confirming this conclusion, genetic findings made since their talk suggest that human and chimp ancestors went on interbreeding for more than a million years after their original split.) Most biologists see the process as beginning when some group from a particular species somehow gets separated from the main body. This development sets in motion a cascade of changes.

First of all, the smaller population doesn't contain all the genetic material of the species as a whole. It's a skewed sample, so that soon the two populations are going to look recognizably different. When that happens, members of the smaller group will prefer to mate with other members of that group, even if the two groups come back into contact and regain access to each other.

Next, as the two populations continue to diverge, the new one may find itself able to exploit food sources that the older population couldn't (and/or, conversely, unable to exploit food sources that the older population could). This outcome has the advantage that both groups may subsequently be able to share the same territory and still avoid expensive and wasteful competition for the same resources.

Finally, certain changes (perhaps physical changes in the sex organs, differences in the number of chromosomes, or some other cause) mean that members of the new species can now no longer mate productively with members of the old species. The onset of this stage is the most generally accepted threshold for speciation, even though universally recognized species such as lions and tigers, or horses and donkeys, can still produce viable, and very occasionally even fertile, offspring. (Maybe even humans and chimps could; for a rundown on the most determined attempt to date, Google Ilya Ivanov.) But the point Foley and Lahr were making is that there's no point at which you could say, "Last year/century/millennium this wasn't a new species, but now it is." As with so many natural processes, there's no point at which you can draw a nonarbitrary line, yet when the process began there was only one species, and now there are two.

So macroevolution is only microevolution that keeps on trucking in a new direction. It's just illogical to believe in one and not the other.

However, this isn't the only way speciation can come about. Speciation can come about through niche construction, and it's highly likely that this was true for at least some hominid speciations.

SPECIATION VIA NICHE CONSTRUCTION

Consider the first three of the six successive niches of our ancestors listed in chapter 6: omnivore (australopithecines), omnivore plus bone-cracker (late australopithecine, early *Homo*), omnivore plus preferential meat-eater (later *Homo*, probably starting with *ergaster* or *erectus*). None of these niches involve differences of place, climatic zone, or whatever—there's no highland man, coastal man, cold-weather man, or the like. All the species concerned were principally distinguished not by where they lived or what kind of climate they needed, or even by their physical form—all were more or less bipedal, all retained whole suites of apelike characteristics. *Ergaster* and *erectus* may have been quite a bit bigger than their predecessors, but that was all. They were distinguished from one another mostly by the way they made their living—what they ate and how they got it.

If you look at a map showing archaeological sites where remains and artifacts of human ancestors have been found, you'll see sites linked to different species clustering together in the same places. That proves nothing, of course, since the same sites could have been used by different species in different epochs. But there's at least one piece of evidence that suggests they shared the same landscapes at the same time.

It was long believed that *habilis* was the direct ancestor of *erectus*, the one evolving into the other in the good old no-more-apes fashion. However, recent discoveries have shown that the two species overlapped by as much as half a million years. And according to Maeve Leakey, head of the team responsible for these discoveries, "The fact that they stayed separate as individual species for a long time suggests that they had their own ecological niches, thus avoiding direct competition."

That's because niche construction can drive speciation equally as well as geographical separation. It just changes the order of the Foley-Lahr stages in speciation. The change of resource selection comes first. We can imagine some protohumans persisting with the old bone-cracking, marrow-extracting routines, while others chose the more dangerous route of exploiting the first fruits of technological advance—the flakes that were created as a by-product in making bone-splitting tools—and using this technology to challenge the big beasts for the meat of large dead herbivores.

The most plausible hypothesis is that the group ancestral to *erectus* split off from *habilis* (or perhaps had a quite distinct ancestry) by going from a catchment strategy based on extracting bone marrow to a territorial strategy based on exploiting carcasses of large herbivores. Such a novel lifestyle would select for both behavioral changes and physical changes. *Erectus* would have had to acquire a rangier build, essential for covering the longer distances that the new lifestyle would demand, and *erectus* was indeed taller and leaner than *habilis*. Doubtless other changes occurred that don't show in the fossil record, such as an ability to withstand thirst, and improved throwing skills. (You'll see why in a moment.) Even if *erectus* did branch from *habilis*, *erectus* females wouldn't want children by *habilis* males anymore. The two species might well coexist in the same territory, using it at different times for different purposes.

What could have set the process in motion?

We can't know, of course. Maybe among the bone-cracking marrow eaters there arose some Stone Age Einstein. (They must have varied in intelligence, just as we do.) Maybe this never-to-be-known genius discovered that the flakes from a bone-chopper could cut hide—how, we'll never know, short of time travel. Maybe he cut himself by accident with a particularly sharp flake. (No, I'm not suggesting that the first word was "Ouch!") If so, perhaps the memory floated to the surface on one of those rare days when, in the course of searching for bones, his band saw, in the distance, a large dead mammal with prospective feasters already congregating around it.

From time to time our remote ancestors must have witnessed such sights, and been fascinated by them. They would have clustered together, staring from a safe distance. All that meat! None of it for them! Why should they have to wait for the other beasts' leavings?

That time, or maybe another, perhaps they approach more closely. They and the scavengers regard one another with the usual interspecies suspicion. Tails are swished by those that have them. The protohumans watch intently, reading the motions of the beasts for the least sign of possible aggression. When none comes, two or three of the younger ones suddenly run forward, briefly mount the hill of dead flesh, and just as quickly race back to their companions—a dare, a show-off for the females. But one happens to have a sharp flake in his hand and as he retreats he stoops suddenly and takes a swipe at the hide under his feet, and the blade goes in and comes out again.

A just-so story, of course. We can't know whether anything like this really happened, or even if it did, how often such scenes took place before familiarity bred courage and the first attempt on a megacarcass was made. If things happened that way, the result of that attempt must have been successful enough to be repeated, and once the strategy was established, everything had to change.

For the new strategy unleashed a cascade of changes.

Those changes may well have been more rapid than the changes brought about by geographic separation. It seems reasonable to suppose that changes driven by animals' own purposive actions would move faster than changes due to genetic drift or even to changed selective pressures. It was precisely such sequences of rapid change, followed by long periods of apparent evolutionary stasis, that gave rise to the theory of punctuated equilibrium.

WHY EQUILIBRIA GET PUNCTUATED

For decades this theory has remained controversial. Now, as with so many things, niche construction theory puts a new light on it by supplying an explanatory mechanism.

It is, and remains no matter what anyone says, a fact of life that most species appear rather suddenly in the fossil record and thereafter change relatively little, if at all, until they go extinct. This fact was what led Stephen Jay Gould and Niles Eldredge to launch, in 1972, their theory of punctuated equilibrium—that evolution moves in spurts interrupted by long periods of stasis. I call it a theory because everyone else does, but in fact, if it's even a theory at all, it's one at the lowest level of theories— a descriptive generalization that does no more than summarize a set of facts. Real theories at least try to explain what they describe. But all the brouhaha pro and con punctuated equilibrium was about whether it really happened or not. No one seemed to notice that no cause for it had ever been suggested by Gould or Eldredge or by anyone else.

Indeed, the alternative to punctuated equilibrium, put forward by Richard Dawkins, didn't have an explanatory mechanism either. Dawkins declared himself in favor of what he called "variable speedism." Sometimes evolution went very fast, sometimes slower, sometimes very very slow, so slow you might think it had stopped until you remembered your dogma: like Monty Python's Norwegian parrot, evolution hadn't ceased to function, it was just taking a nap. Evolution *never* stops!

The trouble was, Dawkins didn't have anything that would explain why the speed of evolution varied from one time to another. But any theory worth its salt that sets out to describe a set of effects is expected also to provide a mechanism sufficient to cause those effects. For example, Alfred Wegener's theory of continental drift made perfect sense even in the absence of any mechanism, but everybody in the business rejected it. (It didn't help that Wegener was a meteorologist, not a geologist.) His theory was nonsense! Have you ever known a continent that drifted? What power on earth or in heaven could make it drift?

Wegener's problem was that he never came up with a mechanism to explain why continents drifted, just as Gould never came up with a mechanism to explain why evolution should alternate between rapid change and stasis. Then plate tectonics was discovered, and everyone suddenly saw that continents just couldn't help drifting.

In the same way that plate tectonics explained continental drift, niche construction explains the otherwise inexplicable stop-go-stop of evolution. A species goes merrily along its way, happily settled in its old niche. Then something in the environment changes; survival demands that a new niche be constructed, real fast. But once that niche has been constructed, when new species and new niche fit like a hand in a glove, what are you going to do? Expand the hand and burst the glove? You stay the way you were, as long as the niche lasts.

Niche construction theory also explains why, since the last common ancestor of humans and apes, there have been so many speciations in our line and so few in the ape line. The ape branch lived in an unchanging environment and stayed happy in the niches it already had. Our branch was forced, at first, and chose, later, as its capacities broadened through successive constructions, to construct more and more new niches. That's why it changed so much and so fast (something else that, before niche construction theory, nobody could explain). Successive niche constructions meant we could evolve in place, without waiting for geographic separation to trigger the speciation process. The process of niche construction was what drove our successive speciations and made us what we are.

But between construction jobs there were long spells of unemployment. That's why our forebears used the same old hand ax for a million years.

THE OPTIMAL FORAGING STRATEGY

Let's now look at this particular speciation, the one that turned bone-crackers into hide-cutters, a little more closely.

The problem is that niche-constructed speciations are triggered by changes in behavior, and behaviors don't fossilize. All we can do is infer them from bones and tools, plus what we know of the habitat—climate, terrain, vegetation, and so on.

The habitat dictated foraging behaviors. It imposed two constraints, constraints that pulled in opposite directions.

On the one hand, the risk of predation called for an increase in the size of foraging groups, for protection. We see this effect in other terrestrial primates, such as the various species of baboon that travel everywhere in large groups and roost at night in still larger ones. The smaller the group, the greater the risk of predation. To forage singly or in pairs would have been even more disastrous than foraging in small groups.

On the other hand, the widely scattered and unpredictable nature of food sources made large-group foraging the least efficient mode for our ancestors. True, they were omnivores like baboons, but they couldn't digest grass, and baboons can. That means baboons can usually find enough to eat in a relatively small area; human ancestors couldn't. Suppose that initially they tried to forage in a large band, but found that even the largest area they could cover in a day supplied only enough food for a fraction of the band. They would then have had to break up into smaller groups.

Some balance had to be struck between the pressure to increase group size (predation) and the pressure to shrink it (scarcity of food sources).

The balance would have been different for low-end scavengers and high-end scavengers. In the absence of hard evidence, let's see how far logic will take us in determining what foraging strategies were best.

Consider range size for the two species. To be more specific, consider day-range size, since there is no necessary connection between the area a group may cover over the course of a year and the ground they may be forced to cover in a single day, if they are to find food enough to survive.

Now consider the relative distributions of bones and recently dead megafauna. Bones last indefinitely; in the absence of any other species of bone-marrow extractors, bones that had lain for months or even a year or so could potentially be exploited. In contrast, the meat of dead megafauna lasts a few days at most. It follows that, on any given day, there will be, in any given area, more bones than there are dead megafauna.

This is what makes catchment scavenging possible. A smaller area is required for it, hence a smaller day-range. If animals can subsist on a smaller day-range, they can afford to increase their group size and thus maximize safety from predation. Baboons with their grass went this route; we may reasonably assume that bone-cracking hominids did the same. But high-end scavenging demands a much larger day-range, for two reasons. First, hominids have to cover a much wider area if they are to locate large carcasses. Second, if they fail to find a carcass (as may have happened on a large majority of days) they have to revert to omnivory to keep alive, and even for an omnivore, savanna food sources are far from plentiful.

Accordingly, it would make no sense for a large group to forage together. Most days they wouldn't get enough for all of them to eat. The only strategy would be to break up into smaller groups. For instance, suppose there was a foraging band of forty individuals that might meet with other bands at night for added protection. If the first band split into groups of, say, eight individuals, it could cover five times as much ground in the same time.

We have no hard evidence that our ancestors foraged in this way, and it's difficult to see how there could be any, or even what hard evidence might consist of, in this case. But the balance of probabilities is strongly in favor of the kind of strategy I've described. It was, after all, only a version of the fission-fusion foraging strategy common among primate species generally. And it sets the stage for a unique adaptation, one that emerges quite naturally from the situation our ancestors found themselves in, once they'd discovered that tools they could easily manufacture could turn them from humble gatherers of the leavings of others into active competitors with the most savage among the beasts.

FASHIONS IN FORAGING

It's ironic, really, when you consider the history of our prehistory, how it's been colored by swings in cultural fashion. In the still male-dominated seventies, man the hunter reigned supreme. The fact that, until well under a million years ago, we didn't have the weapons to hunt big game with was studiously ignored. Then with feminism came woman the gatherer, who supplied the bulk of the diet in terms of fruits, berries, edible veggies, and the like. The fact that, in a Pleistocene savanna, you'd never get enough of these foods to keep you alive was, again, studiously ignored. Nobody ever lets facts get in the way of a culturally appropriate theory.

Meantime, man the hunter had been demoted to man the scavenger. Many men were upset. What a comedown for those prehistoric heroes! No wonder that when both dentition and arguments from gut size indicated that, from around two million years ago, meat formed a significant part of hominid diet, man the hunter came surging in again on a macho backlash.

What were supposed to be the options for paleontologists were starkly put by Craig Stanford, chair of anthropology at the University of Southern California: "Did bands of early humans courageously attack and slaughter large and dangerous game, or did they nervously creep up to decomposing, nearly stripped carcasses to glean a few scraps of meat and fat?"

Well, neither. Those bands engaged in what has been called "aggressive scavenging" or "power scavenging," a third option that Stanford seems to have been unaware of. And here's where the irony comes in. These protohumans were engaging in an activity far more risky, requiring far more macho hardihood than merely (however courageously) hunting big game. True, the latter would have put them in competition with major carnivores, but not normally in direct combat with them. Power scavenging did just that.

And, in a final ironic twist, women most likely did this macho thing too.

We've finally come to the part where I said, at the end of the last chapter, that I'd have to go past the bounds of the certainly known and the all-but-certainly inferable. I could avoid that; I could give you a prosaic summary of how things might have gone, with all the "perhaps"es and "probably"s and "could have"s that academic modesty dictates. Instead I want you to live these moments—I want you to imagine yourselves, at this most critical juncture in our evolution, out there with the forerunners of humanity. I want you, in so far as that's possible, to experience it as they did.

The magic moment approaches

Imagine that we're together, you and I, in this small group, eight of us, drawing breath in the scant shade cast by a thorny tree. Apart from a

small rabbit-sized creature, a few lizards, and a handful or two of withered figs that we squabbled over, we haven't eaten today and it's getting
on for noon. As we sit, briefly, we scan the rolling patterns the wind
makes in the tawny grasses, and the white-on-white sky where motionless streaks of cirrus fail to impede the burning sun, watching for any
sign that there might be food out there.

Suddenly one of us lets out a shout and stands erect, pointing. There
to the west, not too far away, a circling vulture has come into view, followed by another, then a third. We begin to jabber excitedly, pointing
not only toward the sky but to one another, and gesturing. Then we
begin to move.

There's a hill a mile or so away, low, but it should be high enough
for us to see what the vultures can see. We move toward it at a steady,
loping run. The sun beats down on us and we're starting to get thirsty,
but that's par for the course. We can run like that all day if we have to,
not too fast, a steady pace that eats the miles. In perhaps ten or fifteen
minutes—minutes? Who knows what they are?—we reach the summit
of the hill, lie down, and begin to crawl. Parting the grasses, we look
down, and there . . .

There, in a patch of marshy ground, lies the corpse of a huge
deinotherium, a prehistoric elephant. Its hide is intact still, but other
scavengers have arrived already—lionlike or tigerlike creatures, some
larger than those of modern times. A pack of big protohyenas prowls the
perimeter of the scene. The vultures drift on the wind, tightening their
circles. One or two of them land, taking off quickly as a saber-tooth
lunges at them. One lands near us, stares curiously at us with limpid and
strangely innocent blue eyes.

We look at one another. We break and run, you one way, I another.
Nobody tells us which way to go. If one goes one way, another goes a
different way. There's nothing so few of us can do in the present situation. We need numbers, as many as we can get. And we need them now.

How do you get prelingual primates to do stuff?

When I give talks on language evolution, I often start by asking my
audience to imagine that, instead of a hundred or two members of our

own species, the hall is filled with similar numbers from any other primate species. "You will sit here," I tell them, "unless you have urgent business or are outraged by something I say, for the next fifty minutes or so, and you will not move or utter a sound until I finish speaking. What do you suppose would be your chances of getting an equivalent number of apes to stay like that for fifty seconds, let alone minutes?" The answer, of course, is zero.

Language is, among many other things, an unparalleled instrument of social control. There's no coercion involved; I can't have you arrested if you leave, or talk, though if you start throwing things security will probably come in. Cultural norms and expectations take care of it all. But without language, those norms and expectations would not exist.

The biggest problem facing our ancestors as they ran in search of recruits was therefore that of getting those recruits to act as a unit. They had somehow to convince members of the other subgroups that made up their band (and probably also members of nearby groups belonging to other bands) that they should stop whatever they were doing and go after some mysterious target that they could not see, hear, or smell. Why should they? Energy is at a premium in all lifestyles but today's. Until there were machines with power sources external to the animal, to spend energy on anything that didn't have an immediate payoff—whether in terms of calories or mate access or rank within the group—was purely wasteful and potentially suicidal. And primates with smaller brains than these ancestors of ours were fully able and willing to deceive one another. Suppose a group's just discovered, say, a nest of wild bees, and is trying to figure out how best to access the honey with a minimum of stings. Suddenly you come running up to them, gibbering, pointing, and waving wildly. You seem to want them to go with you. What for? Why should they?

It looks like the only way you could get them to go with you would be by telling them what you have found—several days' or even weeks' supply of the most nutritious food around. But you have no language. What can you do?

You are in exactly the situation the ant is in when it finds a large dead caterpillar that it can neither consume alone nor move. Actually you are in a worse situation, because ants are eusocial and you aren't. Ants are all siblings, but with more genes in common than any primate pair has.

On top of which, their brains are too small for them to have minds of their own. Yet they still need to know what they'll be going for, get a sample of it, before they'll move.

You can't give a sample here because of a catch-22—you can't get near the carcass without the help you're trying to recruit. You can't lay a scent trail to it because our much more remote ancestors lost their capacity to deposit and read smells, even in urine, when they took to the trees.

All you can do is imitate whatever species the dead animal belongs to, imitate the sound it makes or the way it moves, or mime some prominent feature of its anatomy.

But that's iconic, you say, and language is symbolic.

Well, even today our vaunted, sophisticated faculty of language isn't ashamed to use iconicity when that's the only way to go. What did the Saramaccan, a group of escaped slaves in Suriname, do when they saw hummingbirds? For most of the animals they met in the tropical rain forests of South America, they could find some analogue among the African animals they'd known back home, so they could give the new animals African names. But there's nothing like a hummingbird in the Old World, and they'd escaped before they'd had time to learn what massa called it. So the word for "hummingbird" in the Saramaccan language is *vumvum*—a conventionalized version of the sound a hummingbird's wings make as it hovers before a flower, a typical piece of iconicity.

Nor do we disdain iconicity when "helping" children to acquire language with words like "bow-wow" and "baa-lamb."

Far more important than what kind of signal was used is the fact that using any kind of signal to indicate an animal carcass at perhaps several miles distance that you'd seen some hours ago would be the first clear case of displacement outside the hymenoptera. Some who write about language evolution make far too much of arbitrariness—the fact that, in today's languages, words hardly ever look or sound like the things they refer to. But the same is true of many ACS signals—a large majority of alarm calls, for example. On the other hand, outside of bees and ants, no ACS signal achieves displacement.

So the real breakthrough into language had to be displacement, rather than arbitrariness. Whatever achieved displacement would do the trick. And iconic signs would work, because in that particular context, with us waving and screaming and pointing "Thataway!," the elephant noise or the hippo noise or whatever it was could have only one meaning: dead megabeast, food for the taking only a short march away.

A SPECIES DOES WHAT IT HAS TO DO

So, back on the savanna, we're recruiting everyone we can.

Including females? You bet. Why not?

The main reason why not is the back-projection fallacy. Among the few hunter-gatherers left in the modern world, men hunt and women gather. Among chimpanzees, females do hunt occasionally but it's mostly the males that do it. So, the easy way to reconstruct our past is to draw a straight line from chimps to humans and assume that our ancestors slavishly followed that line. If females hunted a little in the beginning and don't hunt at all now, women's hunting must have decreased smoothly and gradually through all of the intervening millennia.

This, of course, is just one aspect of the basic ladder model of evolution that, no matter how often it's dissed and dismissed, continues its subterranean life in the thinking even of many who would overtly disavow it. Everything's a precursor of something else, more often than not of us. Everything is still headed for the pinnacle, and we're it.

But in fact, whatever living organisms do is determined by the circumstances they're in, not what some Designer designs for them, or even what some mysterious long-term evolutionary trend tells them to do. There's a saying among Scottish country dancers that if you're not sure of the next figure, "The music will tell you what to do." It's totally false, of course. But in evolution, the niche will tell you what to do, and that's gospel, you'll do it, because your genes are more flexible than you think and can express themselves in a wide variety of ways.

And in any case, we're not talking about hunting. We're talking about high-end scavenging, power scavenging, and scavenging of any kind is something apes hardly ever do—certainly never on the scale our ancestors did it. Power scavenging requires whatever strategies are appropriate for it: in this case, big numbers. And to go in without the women, at least all the women who aren't currently pregnant or minding babies and toddlers, would be like tying one hand behind our back.

Imagine we've now alerted all forty members of our group and they in turn have alerted a few dozen members of other neighboring groups. We're heading back, now, close to a hundred of us converging on the site. And as we go, we're picking up flakes and hand axes.

Something that has long puzzled paleontologists is the enormous number of hand axes that have been discovered. No matter what they were used for, there seem to be far more than would ever be needed.

Why were there so many? Why are they found scattered over the landscape, and why do so many of them show so few signs that they were ever used? The conventional wisdom says they were a chopping tool, used mainly in butchery. Some believe that they were projectiles used in hunting. Some think they were a form of sexual display, made to impress females with their maker's skill. Of course they could have been all these things. But why are there so many, and why are they so little used, and why are they found in so many places?

Suppose one of their principal uses was to drive off the competition at large-carcass sites?

You don't know where the next large carcass would turn up, so you scatter some over the territory and dump others in strategically located caches. Whenever you have spare time you make more and just throw them down or carry them where they might be needed. Then when the call comes you can pick up a half dozen or so, carry four or five in your left hand and the crook of your arm and one in the right hand ready for use. If fifty of you do that, that's near enough three hundred projectiles—aerodynamically functional ones, too, as William Calvin pointed out. He thought they were used primarily in hunting, and from time to time they probably were, but they were also sharp and heavy enough to do serious damage to competing scavengers.

It's the men that do this. The women pick up a single hand ax each and a sharp flint flake or two. They're going to be doing the butchering.

What? Women butchers? But that's a man's job!

Not necessarily. Look at the logic of it. Men are expendable; any man can make a woman fertile. Women are the womb-carriers, the future. If I get killed, my genes will go on in you and your babies. So I'll fight off the beasts while you butcher—it's the only option that makes evolutionary sense.

The site's coming into view now, and nothing much seems to have changed, except there's maybe even more competition than before. The afternoon is already well advanced. By dark we'll be done or we'll be dead.

It's showtime.

We men go out front, a rough semicircle with the women inside it, more men than women because some women are tied down with kids, back in one of our refuges. Beasts on the nearer side of the carcass stir restlessly, start growling. They still haven't gotten used to this new species that does stuff no species did before. They're not sure how to handle us. We start screaming, in unison. Then we start throwing.

The stones aren't big enough to kill large animals, but they're big enough, when propelled with all the force of an arm and shoulder, to knock out an eye or break a big carnivore's leg. A half-blind carnivore or one with a broken leg probably won't live long. It will be weeks, if ever, before a break heals enough for the animal to hunt again. Chances of infection are high. There are no thought bubbles over the carnivores' heads, but inside them, unconscious cost-benefit analyses are going on. They start to back off, snarling. Except one, enraged by a hit, who hurls himself at us.

Two of us go down. The saber-tooth has a grip on the neck of one. It twists him and disembowels him with one rake of its hind claws. We pile on, throwing from shorter and shorter ranges, darting in and leaping back. The saber-tooth reels, recoils, drags itself off with both back legs broken. By now we're all around the carcass—the women are climbing it, they're cutting, blood and lymph oozing from the cuts, and the beasts go wild at the smell. Three more of us go down in their first charge, the three who weren't quick enough backing away from them. At least one won't get up again, but several of the beasts now are limping or have bleeding cuts. The saber-tooth we hammered earlier has stopped and sunk down, hurt worse than we thought, and in a lightning turnaround the other beasts are onto it, working out their frustration and hunger, tearing it apart.

Now it's a standoff. Every time they make a move toward us we start screaming and throwing again. And the women are working fast, going for the best meat, hacking out chunks, smashing through bone where they have to, piling the chunks on the ground beneath, ready for transport.

The trick is, don't be too greedy; leave enough so the beasts won't follow us when we leave.

There's no particular order to it, we're not that organized yet—order will have to wait on language. A woman hefts a big hunk of meat and starts running with it. Others follow. We men start backing up. The beasts are reenergized. Some of them start to come after us. We run, throw, turn, run, throw again. We're almost out of stones. The other beasts are scaling the carcass, plunging into the clefts we've made in it, ripping out dinner. The ones following us slow down, their cost-benefit analyses still ticking. Before them is meat they'd have to fight for and might not get even then; behind them is meat that won't run but won't be there for long if they don't hurry back. No contest. By the end of the first mile, we're on our own.

We have food enough for all of us, several days' worth at least.

And we got all of that with just the smallest possible bit of language!

Why language had to start this way

The scenario I've just described is not original (except maybe for the way I told it). A number of paleontologists have described aspects of it, even down to the need for recruitment. For instance, James O'Connell, chair of anthropology at the University of Utah, and his colleagues wrote a nice recruitment scenario into a paper that (despite optimal foraging theory, despite evidence from guts and teeth, despite the fact that traditional primate foods were vanishingly scarce at the high point of savanna expansion) still insisted that meat formed only a small part of our ancestors' diet. In other words, even opponents of large-scale meat-eating have to accept the logic of this situation:

"Neither would transport of parts [of dead megafauna] to 'central places' be indicated . . . ; individuals or groups *may simply have called attention* to any carcass they encountered or acquired, just as do modern hunters . . . If the carcass had not yet been taken, *the crowd so drawn* could have done so, then consumed it on or near the spot, again just as modern hunters sometimes do" (my italics).

O'Connell wasn't into language evolution; he was looking at foraging behavior and diet, which makes his endorsement of the key ingredient in the forging of language all the more compelling.

But the question we now have to deal with is whether what I've proposed meets all the conditions listed in the first chapter—the conditions that any adequate theory of language evolution has to satisfy. Let's look at each condition in turn.

- The selective pressure had to be strong.

This condition pretty well narrows things down to the only two pressures that determine whether life will go on at all: sex and subsistence. The pressure here involves subsistence, and it operated on a species whose existence as a savanna forager was highly precarious. If our ancestors hadn't stumbled into high-end scavenging, they'd probably have shared the fate of every other previous and contemporary species in our branch of the evolutionary bush, certainly that of the bone-crunching species *garhi* and *habilis*. They'd have gone extinct; you and I would never have existed. Think of a stronger pressure than that.

- The selective pressure had to be unique.

Since no other species above the hymenoptera (excluding maybe ravens) has taken even a step as basic as the one I just described in the direction of language, we can assume that no other species experienced this particular pressure. And indeed, outside those I just mentioned, I know of no other species that has faced a subsistence problem that could be solved only through recruitment.

- The very first use of language had to be fully functional.

Well, you could hardly have anything more functional than what I've proposed here. One word or sign, plus a gesture or two, would trigger a series of events with profound consequences for the immediate future of the group that produced it. In contrast, virtually all the other proposals for language evolution require a minimum of several units (protowords and/or protosigns), more likely dozens or even hundreds, to achieve any result at all. If the first one or three or five protolanguage signs didn't have a substantial payoff, no one would have bothered to invent any more.

- The theory mustn't conflict with anything in the ecology of ancestral species.

I don't think it does. What more can I say? Our prehistory is a con-tentious field—how could it be otherwise when so many things we'd like to know remain unknown, and so many people from so many dis-ciplines try to work in it, each with their own background and their own agenda? Who knows, tomorrow may bring some discovery that will change everything. But don't hold your breath.

- The theory must explain why cheap signals should be be-lieved.

To make a noise like an elephant is not very energy-expensive. Why should anyone believe the message that it carries, when so little effort has gone into producing it?

Well, because of rapid confirmation (or disconfirmation). In the case of most of the kinds of message other theories propose for early lan-guage—gossip, politicking, sexual advertising, or whatever—it's hard, maybe even impossible, for the recipient to check whether the message is an honest one or simply part of the conning, the fakery, the exagger-ation of one's own talents to which all primates, and perhaps we more than any, are prone. This case is quite different. Either there's a dead deinotherium over the hill, or there isn't. In a couple of hours you'll know. If anyone's been foolhardy enough to fake a message, you can kick his or her ass. If not, there's a bonanza of food there for you, enough incentive to make you a believer when the call comes again.

- Finally, the theory must overcome primate selfishness.

One of the big problems facing any exchange of information is, why exchange any information at all? Why tell anyone something that might give them an advantage over you? Why not keep any useful information for yourself, and exploit it for your own benefit?

Exchanging information about the scavenging of megafauna is one case that overcomes this problem. If I don't tell others about the dead deinotherium, there's no benefit. I can't exploit it for myself alone. I get the benefit only if I can persuade others to help me, and I can only get others to help me by giving them information. If we cooperate, we all gain; if not, we all lose.

No previous theory of language evolution satisfies all these conditions.

But that's still not the best part of the story. The best part of the story is that it gives us cooperation for free.

Human cooperation has long been a puzzle for anthropologists. Robert Boyd and Peter Richerson put it like this:

> Our Miocene primate ancestors presumably cooperated only in small groups mainly made up of relatives like contemporary non-human primates . . . Over the next 5 to 10 million years something happened that caused humans to cooperate in large groups. The puzzle is: what caused this radical divergence from the behavior of other social mammals? Did some unusual evolutionary circumstance cause humans to be less selfish than other creatures?

But, like all anthropologists before them, they don't seem able to think of any "unusual evolutionary circumstances." No surprise there: the straight-line theory of human development reigns supreme in anthropology as in linguistics; apes supposedly morphed smoothly and straightforwardly into humans with never a jink or a detour. Accordingly, old and inadequate explanations are constantly recycled: cooperation must have sprung from an extension of reciprocal altruism, from special effects of culture or language, or (last resort of the baffled) from some mixture of the above with other, unnamed factors.

As you may have guessed by now, nobody's quicker than I to attribute anything uniquely human to language. But this is one case in which language won't cut it. Imagine taking any primate species and giving them language, but not cooperation. All you'd get would be a species of screaming wannabe bosses and back-talking don't-wannabe subordinates.

In almost all other species, including primates and other human ancestors, subsistence could be obtained without non-kin cooperation. Foraging, gathering, even hunting could be carried on by individuals or small kin groups. Only a species that came to depend (not completely, of course, but substantially) on accessing giant carcasses would have been obliged to recruit non-kin—obliged, because if non-kin did not cooperate with one another, nobody got anything. And only a long period (probably hundreds of thousands of years) of such activity would have been sufficient for the drive to cooperate to become, in humans, almost as strong as the drive to compete.

On the face of it, recruitment for carcass-exploitation could, initially, have done little more than crack the prison walls of the here and now between which all other primates, for all their intelligence, were confined. But a small crack can have huge consequences. One of the basic tenets of chaos theory is that "small variations of the initial condition of a nonlinear dynamical system may produce large variations in the long-term behavior of the system." The behavior of our power-scavenging ancestors was surely a nonlinear dynamical system. And, as the closing chapters of this book will show, the creation of protowords may have been enough, alone, in and of itself, to trigger the "large variations in the long-term behavior" of that system that would eventually give us full language, human cognition, and (almost) unlimited power over earth and all its other species.

However, there is an alternative story.

In all fairness to you, I must tell it. It comes from the person who is, in the opinion of many, the greatest living linguist, even the greatest linguist who's ever lived, and it was first published in a source with the highest reputation for scientific accuracy. So before I finish my story, let's look at Noam Chomsky's take on language evolution.

9

THE CHALLENGE
FROM CHOMSKY

A DISCLAIMER

Before I start to deal with Chomsky's theory of language evolution, one thing has to be made clear.

There is a widespread industry in the behavioral sciences that you might call "vulgar anti-Chomskyanism." I guess it's just what happens to everyone who gets to be top of a heap; it's great for the ego if you can take them down. But beyond that, Chomsky in particular has been the target of vicious criticism because he is seen to embody one side of a dire rift in modern thinking—the rift between people who (like Chomsky) believe human nature is largely determined by biological factors and people who believe human nature is largely determined by human culture, which in turn has largely broken free from biological constraints. This, like any other scientific debate that involves our species, generates far more heat than light. Luckily for Chomsky, few on the opposite side are as smart as he is. Most of them simply misunderstand him, willfully in some cases. And as far as the rest are concerned, he's a master polemicist; he can more than hold his own with them.

So let me, here and now, dissociate myself from vulgar anti-Chomskyanism. To the contrary, while he's wearing his syntactician's hat I have nothing but respect and admiration for Chomsky. He's a better syntactician than I am. He did the behavioral sciences the greatest service ever when he trounced B. F. Skinner and Skinner's behaviorism. As a result of his initiatives we've learned things about language that couldn't even have been dreamed of a half century ago. Were it not for Chomsky, I could never have launched, or even conceived, my language bioprogram hypothesis (see my book *Bastard Tongues* if you don't know what that was). Linguistics is far better off from having had Chomsky than it would have been if he'd never existed.

I haven't been entirely uncritical over the years, but my position on most issues hasn't differed that much from his, and our names have even been coupled from time to time as twin demons of innatism (whatever that is), usually by people in opposite camps who didn't know too much about us, but at least once by the person who would later, rather surprisingly, become Chomsky's collaborator in evolutionary studies: Marc Hauser of Harvard.

So the only reason I'm going up against Chomsky now is that I'm convinced that on the subject of how language evolved, if on nothing else, he's completely wrong.

But first, we need some backstory to see where he's coming from.

THE STRANGE HISTORY OF AN IMPROBABLE COLLABORATION

For many years, Chomsky had nothing to say about the evolution of language, except that there was nothing sensible that anyone could say about it. Indeed, he went to some lengths to avoid even discussing the topic, and what little he said seemed aimed at discouraging anyone else from doing so. At a conference in 1975, when asked how language got to be the way it was, he replied, verbatim:

"Well, it seems to me that would be like asking the question, how does the heart get that way? I mean, we don't learn to have arms, we don't learn to have arms rather than wings. What is interesting to me is that the question should be asked. It seems to be a natural question; everyone asks it. And I think we should ask why people ask it."

"Why do you ask that question?" was of course one of the stalling tactics in the notorious ELIZA program that, way back in the sixties, MIT computer scientist Joseph Weizenbaum devised to parody a therapeutic session. (If you'd mentioned your father in the answer, the virtual shrink would have continued, "What else comes to your mind when you think of your father?")

Many people besides the original questioner wondered why someone who saw language as part of human biology should show not the slightest interest in the biological history of language.

Then out of a clear blue sky came the 2002 paper by Marc Hauser, Noam Chomsky, and Tecumseh Fitch that, as mentioned in chapter 7, appeared in the "Science's Compass" section of the prestigious journal *Science.* As its placement in that particular section suggests, this paper—portentously titled "The Faculty of Language: What Is It, Who Has It, and How Did It Evolve?"—set out with the avowed object of guiding students of language evolution away from sterile and unproductive bickering and into more constructive ways of thinking about the topic.

The paper caused a stir, not least because of the previous histories of two of its coauthors. Indeed, the surprise of seeing Hauser and Chomsky on the same page was almost as great as what you'd experience if, on opening a back issue of some political journal, you found a position paper on the Middle East coauthored by Yasser Arafat and Ariel Sharon. For until then Hauser and Chomsky had been on opposite sides of at least two of the most crucial issues that language evolution involved.

One was whether or not language developed out of a prior ACS.

As a biologist, Hauser subscribed unquestioningly to the "modern synthesis" of neo-Darwinism, which saw evolution as resulting mainly, if not solely, from the selection and recombination of genetic diversity. Every trait had, therefore, to have immediate and direct precursors of some kind, and "language, as good a trait as any, would therefore be viewed as a communicative form that evolved from earlier forms."

Chomsky, however, has frequently and vehemently dissociated himself from this view, claiming that "it is almost universally taken for granted that there exists a problem of explaining the 'evolution' of human language from systems of animal communication." However, studies of animal communication only serve to indicate "the extent to which human language appears to be a unique phenomenon, without significant analogue in the animal world." He concluded that "it is quite senseless to raise the problem of explaining the evolution of human language from more primitive systems of communication."

The second issue is the role played by natural selection in language evolution.

Hauser regarded natural selection as the principal driving force in evolution generally and language evolution in particular. Writing of Steven Pinker's model of language evolution by gradual increments, each of which was specifically selected for, he described the model as "powerful and lucid," one that "fits beautifully with the conceptual goals of [Hauser's] book"—referring to his *The Evolution of Communication*. He concluded that "natural selection is the only mechanism that can account for the complex design features of a trait such as language."

Chomsky, on the other hand, repeatedly argued against any role for natural selection in language evolution. Discussing various attributes of language, he claimed that "to attribute this development to 'natural selection' . . . amounts to nothing more than a belief that there is some naturalistic explanation for these phenomena . . . it is not easy even to imagine a course of selection that might have given rise to them."

So what brought these unlikeliest of coreligionaries together? In late March 2002, the Fourth International Conference on Language Evolution was held at Harvard, and Hauser and Chomsky appeared together (with Michael Studdert-Kennedy of Haskins Laboratories, an authority on speech sounds) in a two-hour three-handed roundtable discussion. Obviously there's no point in a roundtable discussion where everyone agrees, so I can only suppose the participants were chosen precisely because they still represented very different approaches to the problem. I was down to present a paper at that conference, but couldn't attend, so I can't tell you what magical chemistry clicked into place. But it must have been powerful stuff, since the contradictions were resolved, the joint paper was written, the publication process was completed, and the finished product showed up in the pages of *Science*, all in less time than it takes to make a baby.

If you weren't familiar with their previous writings, you'd never have dreamed its authors had ever been on opposite sides of anything.

Now, I've nothing against authors changing their minds. In fact, they'd better change them, from time to time; otherwise they must be doing faith-based science. I've changed my mind several times, but each time I've admitted it and explained why I changed. I think that anyone who makes a substantive change of position owes it to colleagues and readers alike to make full disclosure, explaining why the old position was wrong and what new facts, arguments, or discoveries led to the change. Certainly no one has the moral right to practice a cover-up, to behave as if they'd always believed the things they believe now. But the *Science* paper, far from being a modest and collegial explanation of its authors' thought processes, adopted a tone that was condescending and didactic, almost hectoring.

For all its air of entitlement, the paper represented a compromise. In a compromise, you have to give in order to get. What did each of our two authors give and what did they get?

The key move that made the compromise possible was to partition the territory of language. Language was now officially divided into two parts: FLB, the faculty of language (broad sense), and FLN, the faculty of language (narrow sense), which formed a part of FLB. FLB was everything in language except the "internal computation system"— whatever drives syntax—and that, at least as a first approximation, was simply recursion (the capacity to embed one linguistic structure within another of the same kind—one phrase, clause, or sentence inside another). FLN was the only part of FLB that was both (a) limited to humans and (b) specifically dedicated to language. The rest of FLB either had antecedents in other species or, if developed by humans, was initially developed for purposes other than purely linguistic ones.

Now that language had been split in this way, a deal was possible that gave each author at least some of what he wanted.

Hauser was able to maintain his belief in natural selection by locating all (or almost all) the components of language in other species, where they might indeed have been selected for, but where naturally they couldn't have been selected for *as* language components. As long as he no longer claimed that language per se had been selected for, that was okay with Chomsky. Hauser was even offered the tempting bait that perhaps even recursion, Chomsky's syntactic holy-of-holies, had non-human origins too. For Chomsky was willing to accept the possibility that recursion had originally developed in some nonhuman species for some purpose that had nothing to do with language (navigation, numbering, and social relationships were three possibilities that were floated).

Chomsky stopped insisting on the uniqueness of language as a whole, and the degree to which it was separate from the capacities of other species. In return he received confirmation of the special status of recursion as the central mechanism in syntax, and syntax was, of course, what he had always regarded as the most essential component of language. Moreover, if recursion in language could be shown to have come from recursion that had developed in a different species for a different purpose, he could continue to assert that it *hadn't been selected for as a specific linguistic mechanism*, which was what he'd been saying all along. He could therefore continue to claim that language hadn't been selected for qua language; it was just that a whole lot of things selected for other purposes had somehow conspired to produce language.

Unfortunately, as so often happens with compromises, a great deal had to be swept under the rug.

THE HAUSER-CHOMSKY-FITCH PROPOSAL

Perhaps the most surprising thing that went under the rug was human evolution.

Nothing else tells you so clearly how far the Hauser-Chomsky-Fitch position is from the position of this book. The position of this book is that the evolution of language forms part of the evolution of the human species, and that to think of one without the other is like *Hamlet* without the prince of Denmark, or Hamlet's speeches without a word from the king, the queen, the ghost, Ophelia, Polonius, Laertes, or the gravedigger. They're great speeches, but the play's the thing. Yet the paper contained not a single mention of human evolution. None of our ancestral species was referred to, not even by bet-hedging designations like "prehuman" or "protohuman" that the current uncertain state of paleoanthropology has obliged me sometimes to use. No times or locations were specified or even tentatively suggested. I've complained about linguists who write as if what was evolving was language X in species Y on planet Z at time T, but two of the authors here were biologists.

If it came to that, there was surprisingly little about Evolution with a big E. Instead, the approach had much in common with that of Charles Hockett, doyen of linguists in his day (and incidentally a target of scorn for Chomskyites as far as his purely linguistic theories were concerned), who in 1960 broke language down into thirteen properties, all but one or two of which, he claimed, could be found in other species (Hauser had written favorably about Hockett's work in *The Evolution of Communication*). The main difference was that while Hockett was concerned with *properties of* language—arbitrariness, semanticity (the conveyance of specific meaning), cultural transmission, and the like—Hauser, Chomsky, and Fitch were concerned with *mechanisms for* language (or at any rate, mechanisms you could plausibly argue were used in language)—ability to discriminate speech sounds, to operate simple rule systems, to acquire conceptual representations, to imitate vocally, and so on. That at least was an improvement. But both treated language as if it were a grocery list—order all these fine products, put them together, and you've got language.

Basic to both the Hockett and Hauser-Chomsky-Fitch approaches was the same crucial assumption: that the properties or mechanisms they specified amounted to the building blocks of language, from the assembly of which language was actually built. This was a dubious assumption in at least some cases; language evolutionists were being urged to go study things like *vocal tract length in birds and primates,* even though birds are very distantly related to us and primates are notoriously unable to produce speech sounds. But what's really wrong here is not the ingredients of the list; it's the list approach itself.

Let's suppose something I think is quite far from the truth, that some prehuman species happened to have collected the full deck—all the precursors, each stepping-stone, every last one of the Hauser-Chomsky-Fitch potentially linguistic mechanisms. And then what? How did all those mechanisms, some of which had never been used for communication, all come together? What made them come together? Since all these mechanisms also did other stuff, why didn't our prehumans just go on doing that other stuff, living their lives like all other species, never developing language at all?

So, even given a miraculous creature that had somehow managed to amass the full set of precursors, the authors have failed to answer the most crucial question in their title, ". . . and how did [language] evolve?" But it's highly unlikely that such a creature ever existed, because many of the mechanisms appealed to are found in animals that aren't even ancestral to humans. For instance, cotton-top tamarin monkeys have been shown to discriminate speech sounds as accurately as

human infants. But chimpanzee hearing is believed to be rather poor;
it's actually been suggested that, even if they could produce speech
sounds, their inability to distinguish those sounds would have prevented them from acquiring spoken language. So at least some of these
precursors couldn't form part of our genetic heritage—all that's been
shown is that, given an environmental need, species are capable of
developing them.

What I'm proposing here is a species that started with only some of
the prerequisites for language, and developed the rest as it went along
constructing the language niche.

Why do so many people seem to find it so hard to grapple with this
concept? I think a lot of it's due to a misunderstanding of how genes
work. There's a widespread belief that genes are immutable and the
instructions they contain constitute nonnegotiable demands. When I
proposed that a biological program for language was what caused the
similarities among creole languages, people seriously thought that by
finding a couple of creoles that didn't contain the predicted features,
they had irrefutably disproved my proposal. The plasticity of the
genome is one of the most underappreciated facts in science. To begin
to grasp it, just think of this: we have only about twice as many genes as
fruit flies, and at least 10 percent of our genes are the same as fruit fly
genes. What do we have in common, physically or behaviorally, with
fruit flies? Next to nothing. The remit of any gene is extremely broad
and vague, executable in a wide variety of ways. So the way any genome
expresses itself is only partly a function of its individual genes; it's partly
due to the context—what other genes each gene has to interact with—
and partly due to the demands of the particular niche that the gene's
owner happens to be occupying, or constructing. All you can say for
certain is that the genes we share with fruit flies must be doing rather
different jobs in them and us.

HAUSER AND ME

Given the gift of "Science's Compass" to point the way to benighted
language evolutionists, what could have made Hauser and his colleagues choose to point it in the counterproductive direction that they

did? Was it simply the inevitable result of the compromise I described above, or was some other factor at work? Back to some backstory again.

In 1996, the journal *Nature* (the British equivalent of *Science*) asked me to review Hauser's *The Evolution of Communication*. After the review appeared, Marc wrote me an amicable if somewhat reproachful letter in which he wondered why I had spent nearly half of the review talking about language. My reply was twofold. First, I pointed out that if the review editor of *Nature* had asked me to review the book, rather than some specialist in animal communication, or at least a professional biologist, it could only be because that editor saw the language connection as a crucial part of the book. Second, the probable reason why he had chosen a linguist rather than a biologist was that the book, for a general survey of animal communication from an evolutionary perspective, paid an inordinate amount of attention to language.

Human language does, of course, nominally constitute a form of animal communication, albeit at best a very lonely outlier in that class. There was, then, room in a book of its type for maybe a brief chapter summarizing some of the major respects in which language differed from other members of that class. But you wouldn't expect to find—nor would even I, a linguist, ever have dreamed of writing—a book on animal communication that, after the briefest of introductions, plunged into forty pages that dealt analytically and critically with what five linguists and two biologists had to say about the evolution of language, that later devoted twenty-eight pages to a discussion of how human language worked, and that sprinkled its text throughout with comparisons between language and other animal systems, finally closing with another dozen pages on how you might design a novel communicative system that would have at least some of the properties of language.

I'm pretty sure Hauser never intended to suggest that animal communication constitutes some kind of pyramid (or ladder) up which all the animals, in their dumb, stumbling fashion, are trying to climb in order to reach the summit, language. Yet that's what an account like this inevitably brings to mind. And if you're wondering why he gave such an account, as I was, maybe the clue lies in this passage:

An overarching concern in studies of language evolution is with whether particular components of the faculty of language evolved

specifically for human language and therefore . . . are unique to
humans. Logically, the human uniqueness claim must be based
on data indicating an absence of the trait in non-human animals,
and, to be taken seriously, requires a substantial body of relevant
comparative data.

My jaw had dropped on reading this. Why was it "an overarching
concern" to tackle human uniqueness? What on earth had human
uniqueness to do with how language evolved? I myself didn't give a
damn about human uniqueness. I just happen to be a member of a
species that happens to have language, and I want to know how and
why we got it. Every species is unique, by definition, because if it
weren't it would be part of some other species.

This whole business of human uniqueness is a red herring, dragged
in by the culture wars. There are some people who want to prove, for
their own ideological reasons, that practically everything about humans
is unique and others who want to prove, for *their* ideological reasons,
that virtually nothing about humans is unique. And, as always, there
are some wishy-washy people who want a "moderate" compromise
between these positions. Hauser falls relatively near, though definitely
not at, the "virtually nothing" extreme.

My reaction is, to hell with the whole thing. Science should have no
truck with this kind of stuff. Just answer the real questions, and let the
uniqueness chips fall where they may. Because once you start worrying
about "human uniqueness," you inevitably start seeing evolution the
wrong way. You start seeing it as how like (or unlike) other species are
to humans. You start, you can hardly help starting, to see humans as the
standard other organisms are measured by. You're doing exactly what
anti-uniqueness people are supposed not to be doing. You're making
your own species the centerpiece of evolution.

Whereas of course evolution doesn't have a centerpiece, or even a
center. And even if it did, it would look too obviously self-serving to put
ourselves there. After all, all we're doing is trying to find out what happened and how and why it happened in the period between us and the
last common ancestor of chimps and humans.

What the paper was telling language evolutionists to do was quite
unrelated to such concerns, and indeed couldn't have been more effec-

tive if it been specially designed to keep them from pursuing such concerns. It was sending them wild-goose-chasing out into the highways and byways, to conduct all sorts of experiments on all sorts of animals, to determine which of the capacities that contributed to language were found in those animals and which were not. The content of the latter category was, hopefully, zero. Although the authors were careful to include bet-hedging language that could be, and was, pulled out later when they were accused of defining FLN too narrowly, it was clear from the tone of the whole paper that they believed, or at worst hoped, that FLN would turn out to be nothing more than recursion, and that recursion, while apparently absent from the behavior of any other species, was really there in some others, but only used for navigation, or social relationships, or . . . *something*.

And if some other species had recursion, how come it hadn't put recursion together with all its other linguistic precursors and made language, long before we did?

Well . . . perhaps because "recursion in animals represents a modular system designed for a particular function (e.g. navigation) and impenetrable with respect to other systems. During evolution, [this system] may have become penetrable and domain-general . . . This change from domain-specific to domain-general may have been guided by particular selective pressures, unique to our evolutionary past, or as a consequence (by-product) of other kinds of neural re-organization."

Okay, class, explain in simple language how an impenetrable modular system becomes penetrable.

"Either way, these are testable hypotheses . . ."

Just how would you test them?

Six years on, nobody's yet found recursion in another species. Of course, for all I know, somebody's finding it right this minute. But I'm not holding my breath.

Chomsky spells it out

The Hauser-Chomsky-Fitch paper didn't go uncriticized. For years a debate raged, or rather simmered, in the pages of the journal *Cognition* between the authors of the *Science* paper and Steven Pinker and Ray

Jackendoff, two authors who had espoused gradualist models of language evolution. Surprisingly, nobody raised any of the issues I've discussed in this chapter. The debate focused solely on what was the proper content of FLB and FLN—shouldn't X, Y, or Z be in FLN, rather than FLB? In other words, it was purely definitional, ignoring, just as the original paper had, all the deeper issues, all the issues most central to language evolution, and worrying about what was unique to humans and what wasn't, as if this were an issue of substance. Pinker and Jackendoff fought Hauser et al. on the turf the latter had marked out.

Also lost in the shuffle was any debate on what if anything happened between the dawn of language and its full flowering. You'd think Jackendoff, with the nine stages he'd hypothesized for language evolution, or Pinker, with the one-rule-after-another scenario that he and Paul Bloom had laid out in a 1990 paper, might have complained about this lack. For that matter, Hauser, who had written favorably of the protolanguage concept, might have been expected to bring up the issue of exactly how language, once started, had developed. This shows just how far "Science's Compass" had succeeded in directing people away from the real issues in language evolution.

Then it occurred to me that perhaps the omission from that paper of anything between no language and full language had been deliberate, not accidental. Perhaps anything intermediate between no language and full language had to be one of the things that, to keep the Chomsky-Hauser compromise afloat, just had to go.

Was that possible? Could they actually be claiming that language had sprung, fully formed and complete, straight from Jove's brow?

The *Science* paper never explicitly claimed that. And surely it couldn't be claiming that, since any such claim would be about as unbiological and unscientific as you could get. Yet if all other bits of language were there in other species, and only one thing, recursion, was then required to give birth to full language, what other interpretation could you put on it?

Then, at the height of my mystification, like a deus ex machina, Chomsky himself appeared before me to sort it all out.

Chomsky seldom goes to linguistic meetings, seldom goes anywhere where he can't devote equal speaking time to language and politics. However, since he was trying to make up for years of lost time and prevarication, he appeared at a meeting on language evolution held at the Stony Brook campus of the State University of New York in the fall of 2005.

The event was billed as a symposium, one of those Greek thingies where you get to hang with Plato and the boys, drink lots of wine, and explore deep philosophical issues together. Maybe those of us who didn't know him were hoping Chomsky would stick around long enough to quaff a glass, crack a joke, spin a new theory or two. Of course that was not to be. Chomsky is a busy man; he told me himself he spends six hours a day (or night) just answering correspondence, a hundred or more letters at a go. He came one evening, had dinner, gave a talk, and was gone by noon the next day.

However, he did leave us with an unusually clear statement of how he thought language had evolved, to which we now turn.

> In some small group from which we all descend, a rewiring of the brain took place yielding the operation of unbounded Merge, *applying to concepts of the kind I mentioned* . . . The individual so rewired had many advantages: capacities for complex thought, planning, interpretation and so on. The capacity is transmitted to offspring, coming to predominate. *At that stage, there would be an advantage to externalization,* so the capacity might come to be linked as a secondary process to the sensorimotor system for externalization and interaction, including communication . . . It is not easy to imagine an account of human evolution that does not assume at least this much, in one or another form [my italics].

This needs a little parsing.

For instance, "externalization" means "to start talking." All these capacities for complex thought, planning, interpretation, and so on were up and running, and only then did it occur to someone to think, "Hey! Why not use them to chat with?"

Then, when Chomsky mentions "concepts of the kind I mentioned," he is referring to an earlier paragraph in his talk, where he stated:

> There do, however, seem to be some critical differences between human conceptual systems and symbolic systems of other animals. Even the simplest words and concepts of human language and

thought lack the relation to mind-independent entities that has
been reported for animal communication . . . The symbols of human language and thought are quite different . . .

What Chomsky points out here is indeed a profound difference
between humans and other animals, one we've looked at several times in
previous chapters. In animal communication, as we saw, the kind of
"functional reference" you get (if you get any kind of reference at all) is to
"mind-independent entities"—directly to actual things in the real world,
rather than to our concepts of those things, the way in which words refer.
When we use a term like "leopard," we're referring (unless we qualify it in
some way) not to a particular leopard, least of all to one that's there before
our very eyes, but to leopards in general; the statement "Leopards live in
Africa" is not invalidated if a couple of them happen to be living in our
local zoo. But when an animal gives a warning call for "leopard," this
refers to one particular leopard, the one that happens to be right here right
now. So in this paragraph, Chomsky is focusing on properties of human
words, and in the previous paragraph he is saying that human words alone
can be merged with others of their kind to form sentences.

Fine so far; just what I've been saying here, and I've even explained
why it had to be so, why animal signals with their "mind-independent
reference" just couldn't ever, in the nature of things, combine.

But Chomsky is also stating that human words (and the typically
human concepts that underlie those words) are the only kind of unit
that recursion can apply to. So when recursion emerged (through a
"rewiring" of the brain) these concepts must already have been present.
This commits Chomsky to the following sequence of events:

TIME 1: Animals have concepts that won't merge.
TIME 2: Typically human concepts, which will merge, appear.
TIME 3: The brain gets rewired.
TIME 4: Merge appears and starts merging typically human
    concepts.
TIME 5: Capacities for complex thought, planning, etc. develop.
TIME 6: People start talking.

So what Chomsky has done is simply add, to the problem of how
language evolved, a second and at least equally horrendous problem—
how typically human concepts evolved.

"At least equally" is probably an underestimate, for the following reason.

We talk about natural selection selecting from genetic variation. But that's shorthand for more complex processes; it seriously oversimplifies what's happening in evolution. Genetic variation is invisible to natural selection. Natural selection can't pick among genes; it can only pick among physical organisms, selecting those that survive longest and breed most. Did the genes cause some to survive longer and breed more? Well, again, not directly, because what determines long life and successful breeding can only be how physical organisms behave. Sure, a species' range of possible behaviors is determined by genes; if those genes can't make wings, flying is out of the question. However, genes don't determine which of those behaviors animals will produce, and behaviors are the proximate cause of whether an animal survives and breeds or not. Real-world events, the interactions of a physical organism with its environment, are what drive natural selection, not things that happen inside the organism.

But for Chomsky, language had to evolve inside the organism before it could appear outside the organism.

Well, that's okay, you say. Animals that could think and plan for the future, even if they couldn't talk, would surely outperform and outbreed nonthinkers and nonplanners.

True enough. But how did they get able to think and plan for the future if they could only do this by first acquiring human-type concepts and then merging them to produce consciously directed, constructive trains of thought? To say animals got gradually smarter through some kind of feedback process simply won't cut it. It's not a question of getting smarter. It's making a jump from "functional reference" to full symbolic reference; from concepts that can't combine to concepts that can combine. No way can you do that gradually. There are no concepts—in fact there is nothing anywhere—that can a quarter combine, or 65 percent combine. Things can either combine or they can't.

As for "the rewiring of the brain," brains don't rewire themselves for no reason, or because they just feel like a spot of rewiring. Brains rewire themselves, to the extent they do, because things are happening in the outside world, things that give individuals with rewired brains an advantage over individuals without them.

So the question becomes, what would have selected for kinds of concepts and kinds of brains different from those that had served all life-forms well since life began?

Answer: nothing.

A concept is something in the mind. Once it exists, it can affect behavior. Before it existed, it couldn't. All that natural selection can see is behavior. So concepts could only have been visible to natural selection once they existed, once they'd begun to affect behavior. But they couldn't exist until they'd evolved, and they could only evolve if they were selected for.

So human-type concepts couldn't have evolved by themselves. They could only have evolved if some other thing had been selected for, something that *was* visible to natural selection—in other words, some overt behavior that gave an adaptive advantage to those that had it.

But for Chomsky there's nothing. All parts of language, everything necessary for its full flowering, just grew, somehow, in the mind, all by themselves.

When I meet someone whose ideas are different from mine, I always try to think through them, determine why those ideas were chosen, rather than others. To find out where Chomsky's coming from, we need to look at two aspects of his thinking about language that, between them, virtually force him to accept those ideas. Both aspects, when applied to language today, have proven both useful and (probably) right. It's when they're applied to the evolution of language that they create problems.

How can we know what we think until we see what we say?

One of those aspects concerns the balance in language between thought and communication.

Chomsky has always emphasized that language is at least as much a system for structuring and thinking about the world as it is a vehicle for communication. I couldn't agree more. I wrote a whole book saying just that, and exploring the world-creating capacities of language, back in 1990.

But I overestimated somewhat, and Chomsky overestimates a great deal, how much thinking animals could do without language.

The fact that language is by now the main engine of thought doesn't have any implications for its status when it began. That's the fallacy of first use, the idea that whatever a thing started doing will be what it does mostly nowadays—and vice versa, naturally. It was the fallacy of first use that led Robin Dunbar to propose gossip as the engine of language evolution, just because gossip is what (spoken) language is most used for today. If the fallacy of first use were true of computers, they would have been used first for e-mail and Web browsing, and some of us are old enough to remember how big a fallacy that would be.

Certainly, language is now the means by which we structure the world of thought, but it would never have gotten off the ground, never developed into what it is today, and certainly never have raised thought to a new power if it hadn't first entered the real world in the tangible form of communication. As I showed above, only external events can shape internal events, because only external events are visible to natural selection.

Chomsky and his followers, who have never liked natural selection, will say that there are other factors in evolution. They will claim that much of evolution results from mysterious, yet-undiscovered laws of form, laws of development, or other similar forces. Even Fibonacci numbers—the strange sequence of figures that seems to control the branching of plants—have been invoked. But this is sheer hand-waving. We know that natural selection works and we know how it works to cause new evolutionary developments. But while things like laws of form may exercise constraints on such developments, no one I know of has suggested that they can cause them, and nobody has the faintest idea, if those laws do work, how they would work to produce the effects of language.

WITH MY METHOD, YOU CAN LEARN LANGUAGE INSTANTANEOUSLY

The other aspect of Chomsky's thinking about language that has had an unfortunate influence on his ideas about evolution is an idealization: the idealization of instantaneity in the acquisition of language.

Some people who studied how children acquired language were very upset by this. After all, children take at least a couple of years before they become fluent speakers of a language, and since we go on learning new words all our lives, you could even claim the process, far from being instantaneous, is one that never stops.

But think this through more carefully. What do children start with? Meaningless babbling. What do they all, all normal ones anyway, arrive at? Full control of a human language. Is there any difference between them at that stage? Well, some may talk more than others, some nicer, but there's no nonsubjective way to determine any difference in how they talk. Whatever happened along the way, whatever funny things they may have said or had said to them, made no difference to the end result. So the process might just as well have been instantaneous. And if they come equipped, in some yet-to-be-defined sense, with some kind of universal grammar—and there's good reason to believe they do—then if it weren't for the severe limitations in speech production, focused attention, word-learning, general experience, and the like that all infants labor under, they jolly well *might* acquire language instantaneously.

The idealization of instantaneity is a legitimate idealization. Science is built on such.

Unfortunately, Chomsky applies this same idealization of instantaneity to the acquisition of language by the species, and that's a very different kettle of fish.

In a child's acquisition, the faculty of language is already there, has been there for tens of thousands of years, waiting for kids to rediscover it, or perhaps merely switch it on. In the species' acquisition, there was a time when there was no faculty at all, zero, zilch, a time when that faculty had to be built from the ground up. An idealization that works well in describing a state does not necessarily work at all in describing a process. Chomsky's mistake is to treat a state and a process as if they were the same thing. Or it might be more accurate to say that he tries to turn a process into a state. That would be natural enough. States he understands; he's been thinking about them all his life. Processes, that's another matter.

You'd almost think Chomsky regarded the evolution of language as a state rather than a process. He treats the faculty of language as if it were already there, in the species. After all, in both cases, child and species, we know that eventually "unbounded Merge must appear"—it *did* appear, didn't it, in both cases?—so in neither case is there any point in looking for intermediate stages between the absence of Merge and its presence.

After all, all it took for Merge to emerge was "some rewiring of the brain." And there couldn't have been anything you could call a proto-language, because until Merge emerged, there was no way to put words together, and once it had emerged, you were already at language and there was no room for any kind of protolanguage.

I'd realized from Chomsky's talk that I'd been right about the Hauser-Chomsky-Fitch model of language evolution not allowing for any kind of protolanguage. I hadn't realized, until I began to correspond with him by e-mail on this issue, that he didn't believe there *could* be a pro-tolanguage.

In the course of this correspondence, I was therefore startled to encounter the following sentence: "It is a logical truism that protolan-guage either involves a recursive operation or is finite."

"Logical truism"? I had the advantage of coming to language evolu-tion from the study of pidgins and creoles, and the most certain thing I'd derived from that study was the fact that pidgin and creole speakers put words together in different ways. Creole speakers put words together the way everyone else who speaks a full human language puts words together, that is hierarchically, in a treelike structure—schematically, $A + B \rightarrow [A\,B]$, $[A\,B] + C \rightarrow [[A\,B]\,C]$, and so on. That, as I understand it, is the process Chomsky calls Merge. Pidgin speakers, on the other hand, put words together like beads on a string, $A + B + C$, etc., so that, in contrast with Merge, the relationship between A and B is no differ-ent from the relationship between B and C. So I promptly wrote back, "Protolanguage consists of $A + B + C \ldots$ i.e. there is no Merge."

"That's commonly believed," Chomsky equally promptly answered, "but it's an error. A sequence a, b, c . . . that goes on indefinitely is formed by Merge: a, $\{a, b\}$. $\{\{a, b\}\,c\}$ etc. . . . If we complicate the oper-ation Merge by adding the principle of associativity, then we suppress $\{,\}$ and look at it as a, b, c . . ."

Principle of associativity? What had that got to do with language? It's a principle in logic and mathematics, and all it means is that moving or removing brackets in logical formulae doesn't affect their truth value, and moving or removing brackets in additions doesn't affect the sum—

$(1 + 3) + 2$ adds up to 6, just the same as $1 + (3 + 2)$ or even $1 + 3 + 2$. You see, the brackets you sometimes find in logical formulae or sums of addition aren't doing anything serious in the first place. They can be rearranged or dispensed with precisely because they don't make any changes in the relationships between things; there is no meaningful sense in which 3, in $(1 + 3) + 2$, is nearer or more closely connected to 1 than it is to 2.

But that doesn't apply in any shape or form in language, otherwise an [English [language teacher]] would mean the same as an [[English language] teacher], and it doesn't: the first means a teacher of languages who happens to be English, and the second, someone of any nationality who teaches the English language. Change the brackets here, you change the meaning; remove them, you just make the phrase ambiguous. Or take a more famous example: [old [men and women]] versus [[old men] and women]. The differences between such pairs can be spelled out by intonation features: stress on "English" in [English [language teacher]] and on "language" in [[English language] teacher] for instance.

But in protolanguage, for example in an early-stage pidgin, there are no structural relationships among words—only semantic ones, so there's no equivalent way to disambiguate stuff. Moreover, you don't have to go as far as pidgin or protolanguage to find beads-on-a-string joining things together. Merge operates only up to the level of the sentence. Phrases have to be properly merged with phrases, clauses with clauses, but once you get up to sentence level, beads-on-a-string takes over. Paragraphs, pages, chapters, books—there's no limit to the number of sentences you can string together. A finite process? No way.

Grammatical relations, relations created via the Merge process, are found only *within* sentences. There are no grammatical relations *among* sentences. There are no agreement phenomena that link one sentence with another, no sentence serves as subject or object of another, no sentences are Agents or Themes or Goals of other sentences, no noun in one sentence can bind a pronoun in another, no adjective in one sentence can modify a noun in another. Sentences are linked only in terms of discourse coherence, which in turn is determined by semantic and pragmatic, not grammatical, factors. Take a sequence like:

"John is looking after his little sister. The price of copper fell 17 percent overnight. The national vegetable of Wales is the leek."

As a paragraph, this is nonsense. But all the sentences that compose it are fully grammatical and, within themselves, both comprehensible and semantically appropriate. In isolation they're fine; together they're nonsensical, because they bear no relation in terms of topic. But now look at this sequence:

"The price of copper fell 17 percent overnight. A sudden drop had been predicted by many analysts. The recent cease-fire in Central Africa has already resulted in an increased supply."

The exact same conditions apply; the sentences are merely strung together, and, apart from "the," no word appears in more than one of the sentences, so there are no lexical links. On the other hand, to anyone familiar with politics and economics, the paragraph makes perfect sense.

So both processes, Merge and beads-on-a-string, work alongside each other in human language. Since beads-on-a-string is simpler than Merge, it's probably older. If it's older, it's only reasonable to suppose that, at the dawn of language, Merge hadn't yet emerged and beads-on-a-string was all our remote ancestors had. A system without Merge doesn't therefore have to be finite—you can, in principle at least, go on adding beads to a string for as long as you have beads. So Chomsky's "logical truism" is simply false.

Evolution in a drum

So let's put the two evolutionary models side by side:

MINE	CHOMSKY's
TIME 1: Animals have concepts that won't merge.	TIME 1: Animals have concepts that won't merge.
TIME 2: Protohumans start talking.	TIME 2: Typically human concepts, which will merge, appear.
TIME 3: Talking produces typically human concepts.	TIME 3: The brain gets rewired.
TIME 4: Merge appears and starts merging typically human concepts.	TIME 4: Merge appears and starts merging typically human concepts.
TIME 5: The brain maybe gets rewired (plausible but not certain).	TIME 5: Capacities for complex thought, planning, etc. develop.
TIME 6: Capacities for complex thought, planning, etc. develop.	TIME 6: People start talking.

The stages do not differ substantially in their content, but the ordering of the stages is very different. And even this is not the most important difference. The most important difference is that in the first model, one stage drives the next. In that model, once our ancestors started talking, their iconic or indexical signals gradually formed into true symbols through variations in their manner of use (this theme will be developed in the chapters that follow). Merge, a process that does not have to be specially derived, since it arises through the way the brain handles any data, appears as soon as there are units semantically capable of being merged. Use of Merge in both language and thought selects for any development in the brain, whether mutational or epigenetic, that will expedite or automate language and thought processes. The end result of all these processes is a vastly improved thinking machine, not to mention a fast, subtle, and highly flexible language.

One stage driving the next, each new development changing the selective pressure for the next—that's how evolution works. In particular (and more rapidly, because the processes are more focused), that's how niche construction works.

Now in Chomsky's model, the most crucial stages have no motivation whatsoever. Nothing drives them. Typically human concepts pop up out of nowhere. The brain gets rewired for no particular reason. People suddenly start talking, again for no particular reason (just because "there would be an advantage" to it!), and the menial details of *how* they started talking, how, with all those concepts whizzing around in their heads, they got to agree on how to label them, gets swept under the rug.

Chomsky's version of evolution does not intersect anywhere with the realities of the world or the realities of evolution: it's evolution in a drum, a totally abstract, self-encapsulated procedure. Yet it is implicit in a paper published in America's flagship science journal, and coauthored by two biologists. Go figure.

Note that far more is involved here than merely the way in which language evolved. At issue is the whole relationship between language and thought. If I'm to deliver on the second half of the subtitle of this book, I shall have to tackle that. And here again Chomsky and I find ourselves in diametrically opposed positions:

Chomsky believes that human thinking came first and enabled language.

I believe that language came first and enabled human thinking.

One or the other has to be true. It isn't even possible to take some wishy-washy, middle-of-the-road position and say, Well, it's a little of both, they coevolved. Many people might like to take that route. "Coevolution" is a fashionable word nowadays. You only have to murmur it, and throw in "meme" and "self-organization" for good measure, and people will nod knowingly and be filled with respect for you. And of course, once the processes I've talked about were fully established, language and human thought most certainly did coevolve.

But at the beginning, it's a hen-and-egg, horse-and-carriage problem. One had to come first, and there's a logical truism if ever I saw one. So next we'll see which one did, and how it did, and why.

10

MAKING UP OUR MINDS

Let's look on the positive side. Chomsky's position focuses attention on the brain. Granted, the statement that "the brain got rewired" positively bristles with "what?" and "how?" and "why?" questions, and will hardly serve us as a guide. And so far, we've been looking mostly at behavior. But brains and behavior are intimately linked, and we've reached a point from which we can't go much further without taking brains into account.

Regardless of whether language is primarily biological or primarily cultural, the brain has just one way of putting it all together.

What do brains, the brains of all other species, and for a lot of the time ours too, actually do? According to Gary Marcus of New York University, the brain "takes information from the senses, analyzes that information, and translates it into commands that get sent back to the muscles." And that's all that brains were specifically built to do, because that is sufficient for life on earth. It's enough in most cases to keep the brains' owners fed and alive and able to pass on their genes to another generation. Brains weren't made to think about the nature of the universe or the laws that govern it. Or even about our own personal affairs—unless we're actually busy with them at the time.

Brains don't (normally) do what they don't have to do, because brains are energetically expensive. Ours use 20 percent of our energy, though looking at some people you mightn't think so. Purists have got down on folk for comparing brains to computers. All cultures, they sneer, have compared the brain to their own most modern technology—the Greeks to water mills, Victorians to telephone exchanges; it's just a fad. But in fact the purpose of a brain is exactly that of a special kind of computer, an onboard computer, like you have in boats or cars or planes or space stations.

The purpose of an onboard computer is to preserve the homeostasis of whatever it's on board of. It does so by monitoring many conditions, both internal and external; anything from keeping internal temperatures within a narrow range to warning you you're about to bump into something. But its range of behaviors does not include sending you messages it constructed itself, for its own ends. It doesn't have its own ends. It has what's been programmed into it.

Human engineers program onboard computers, but evolution programmed the brain. It programmed the brain for homeostasis, to ensure, as far as it could, that conditions in and around the organism that housed it remained as favorable as possible to that organism's well-being.

As Marcus suggests, the brain does its job in a series of steps, along a one-way trajectory:

- Receive information from senses.
- Send it to be analyzed for identification.
- Choose a course of action based on the analysis.
- Send an order to execute that action.

Thus an odor is detected; the odor is compared with other odors and their possible causes; the odor is determined to be that of a predator, but taking its strength plus prevailing wind conditions into account, most likely a predator at some distance; consequently two messages are sent, one to the muscles—"Freeze!"—one to the attention: "Remain on high alert until further notice."

Note that the action is unidirectional, and without side trips. True, feedback from the animal's own actions following the order may influence subsequent developments, but can only do so by reentering the process at the beginning again, making a closed loop.

Now see what happens when you think even the simplest of thoughts, say, "Roses are red."

- Think of "roses."
- Think of "red."
- Connect the two.

You may, or you may not, have a visual image of a red rose. If you do, you will say, "I think in images." If you don't, you will say, "I think in words." In both cases that's like the sun crossing the sky—not what's really happening at all. There are no images in the brain. There are no words in the brain. All that's there are neurons and their connections and differential rates and strengths of electrochemical impulses. These provide a subjective sense of words and images. The metamorphosis may seem magical but it's no more magical than the "changing colors" of mountains at sunset, likewise produced by processes in your brain.

If we define "thinking" as "any kind of mental computation"—surely the most general and theory-neutral way of defining it—then both of the series of operations I've just described (processing information from the environment and thinking of something such as a rose) can be legitimately called "thinking." But beyond the fact that they're both brain-internal, brain-directed processes, there's hardly anything in common between the two.

It seems reasonable to suppose that the first kind of thinking, the brain's "business as usual," you might say, may best be characterized as "online thinking"—thinking that takes place as a consequence of ways in which the thinking organism is, right at that moment, interacting with objects and events in the outside world. If that is so, then the best way to characterize the "Roses are red" kind of thinking is "offline thinking"—thinking that has no necessary or direct connection with what's happening outside, but that is generated and takes place wholly within the brain. It is worth comparing the two in a little more detail.

The steps in offline thinking are quite different from the steps in online thinking, they do not include any part of those steps, and they work in quite different ways. The online steps are triggered by events outside the organism. The offline steps could be triggered by an outside event but they do not need to be, and most often probably are not. In the online sequence, each step triggers the next. If it didn't, the owner of the brain wouldn't last long—if the odor wasn't sent for analysis, if the analysis didn't trigger action, or if the order, when issued, was not obeyed, you'd soon be cat meat. In the offline sequence, no step is necessarily linked to the next; you may think of "roses" first and then "red," or the other way around, or both simultaneously; it makes no difference. In the online sequence, the last step is usually an instruction to the body, even if that instruction is simply "Do nothing." In the offline sequence, there need not even be a last step. You can think of redness, or roses, or both, without necessarily even putting them together, let alone doing something about it.

Perhaps because of this apparently looser structure, and what seems like the basic simplicity of thoughts like "Roses are red," many people assume that we get them for free, so to speak, just by having a brain. And this feeds into another widespread belief, held even by some who accept language as the prime motor of human intelligence, that thoughts are somehow logically prior to sentences, that language arose to express thoughts, that you first have to think something before you can dress it up in words and send it out into the world. Remember, there are not only no images and no words in the brain, there are no thoughts there either—only a continuous cascade of neural activity, of pulse rates spiking, of impulses going every which way.

For me, the belief that thinking preceded speaking in evolutionary time and precedes it operationally in our daily lives is one of those initially plausible beliefs that when held up to the light and carefully examined can be seen to lack any real foundation in either fact or theory. Indeed, I am going to argue that, until we could talk, we could not even think, "Roses are red." But since you may well be harder than me to convince, let's look at what's actually going on in both human and nonhuman minds and brains.

Nonhuman versus human minds

Start by thinking of the simplest answer to the question, why are other animals trapped in the prison of the here and now?

The simplest answer is, that's all they've got.

They can't communicate about things beyond the here and now because they can't direct their thoughts outside the here and now. And the reason they can't do this is also the reason they can only refer to specific, immediate events. They don't have abstract concepts, and we do.

Be warned, this is heretical stuff. As a trio of fellow heretics (Derek Penn and Daniel Povinelli of the University of Louisiana, and Keith Holyoak of the University of California, Los Angeles) pointed out in a recent article, "Ever since Darwin, the dominant tendency in comparative cognitive psychology has been to emphasize the continuity between human and nonhuman minds and to downplay the differences as 'one of degree and not of kind' (Darwin 1871)."

For example, in asking his readers to imagine themselves at the dawn of language, Jim Hurford, one of the best writers on language evolution, suggests with breezy confidence that they should think what it would be like to be without language "but otherwise cognitively pretty much as you are."

But surely nobody would make such claims without massive evidence?

Right; according to Irene Pepperberg, Alex the parrot's mentor and advocate of a psittacine-inclusive approach to language evolution, "for over 35 years, researchers have been demonstrating through tests both in the field and in the laboratory that the capacities of nonhuman animals to solve complex problems form a continuum with those of humans." Anyone who goes up against this consensus risks being branded as a last-ditch defender of human supremacy, the special, divine creation of "just us."

But is this evidence telling us the plain truth or the truth we'd like to hear—whatever reinforces the current brand of the Darwinian wisdom? Could it perhaps be interpreted in quite a different way?

Before we review the evidence, let me set things up by looking a little more closely at the Irene Pepperberg quotation above: "the capacities of nonhuman animals *to solve complex problems* form a continuum with those of humans" (my italics). But solve them where? In the world or in their heads? That's the crucial question.

What I'm going to claim is that the capacities of nonhuman animals do indeed form a continuum with those of humans *when it comes to solving physical problems in the real world.* Such problems are highly likely to be solved if:

- The animal is highly motivated to solve the problem (that is, if the problem stands in the way of fulfilling some immediate need for food, sex, escape, or anything else that contributes directly to the animal's fitness).
- Most if not all the things necessary for solving the problem are in plain view.
- Anything not in plain view is stored in the animal's episodic memory (the part of memory that, roughly speaking, stores its

owner's experiences in narrative form) and can be triggered by attempting to solve the problem.

Problems of this nature may even be solved faster than a small child or an adult with a low IQ could solve them. (If you've seen the video in which a New Caledonian crow, given a straight piece of wire to get food out of a glass tube, twists the wire into a hook after only a few seconds of fruitless poking, you'll know what I mean, and never again use the demeaning expression "birdbrain.")

The question is not whether animals can solve problems—they obviously can—but whether they have concepts that they can summon at will and manipulate so as to imagine, and thus subsequently produce, novel behaviors.

But mightn't some animals have concepts too?

It's more than likely that a majority of people from the behavioral sciences will at this point if not before cry out, "But *of course* animals have concepts like ours!" To deny this means claiming there's a discontinuity between humans and nonhumans. But there's already one possible discontinuity, that between language and all other forms of communication. One such discontinuity is bad enough—two, and the whole edifice of Darwinian gradualness begins to shake. Or so it (wrongly) seems.

Think this through a little further. Unless we are ready to accept that bacteria have concepts (and where on earth would they store them?) there has to have been, somewhere in the course of evolution, this exact same discontinuity between animals that had concepts like ours and animals that didn't. If we don't place it between humans and nonhumans, where are we going to put it, and why? Given all the things we can do that other animals can't, what likelier place to find that discontinuity than between us and them?

In the final analysis, the issue of whether or not animals have concepts like ours is an empirical one. They either do or they don't. It should be possible, within the next few years, to determine through advanced brain imaging techniques if the difference between us and the rest of nature really is what I've suggested here. In the meantime, we should use any other means available to determine whether any other animals have concepts that they can conjure up at will—that they can start thinking about even if there's nothing currently happening that would directly or indirectly activate those concepts, just as we do. If it turned out that some did have such concepts, we'd have to start looking all over again for why we're so much more creative than other species.

In the meantime, let's look at some of the best evidence for animal concepts.

Back in the seventies, Richard Herrnstein (later to gain notoriety by coauthoring a book he never lived to see published, *The Bell Curve*) carried out a series of experiments with pigeons that psychologists are to this day still trying to explain. He himself was dumbfounded: "How can animals with such powers of classification still seem stupid in some ways?" he asked. Surely, only if their minds work very differently from the way ours work.

By training pigeons on a small sample set, he first got them to reliably distinguish between pictures that had trees in them and pictures that didn't. (They did this by pecking when there was a tree in the picture and not pecking when there wasn't.) Easy, you say; pigeons spend half their lives in trees. So Herrnstein moved them on to pictures of people, and they did equally well. Okay, pigeons have usually seen people; certainly all pigeons in a psych lab have seen people. But *fish*?

When Herrnstein trained his pigeons on a small set of fish pictures and then turned them loose on a much wider collection they'd never seen before, they did pretty much as well as they'd done with trees and people. You can be absolutely certain that none of the pigeons had ever seen a fish or even knew what a fish was. But that didn't stop them from recognizing almost every fish they saw, and knowing when something wasn't a fish.

How did they do it? Did they go by general outlines or specific features or what? Whatever they used, what did it imply about how their minds worked? Thank God we don't have to go there; it's a quagmire. All we have to ask is, does what they did imply they acquired any kind of general concept of fish?

I don't think it does. I think they noted and stored a set of distinct features—doesn't matter what; for our purposes, anything common to fish—and that when a large enough subset of these features was triggered by the picture, they pecked. I don't suppose for a moment that they ever thought about fish once the experiment was over—wondered what they were, imagined what eating one might be like. True, in some cases the training effects lasted a year or more, but that just shows pigeons have good memories; it doesn't show they can form any central notion of fishiness.

So let's go out of the lab into the wild. Let's check out scrub jays.

Scrub jays are a bird species common in the West. They live by collecting seeds that they cache all over the landscape so that when winter comes they'll have a steady food supply. In the course of a season they may make hundreds if not thousands of these caches. And they remember them all, fly back to them unerringly. More than that, they remember which caches have more perishable seeds and which have less perishable ones, and they don't bother going to the first kind in the later months of the winter.

I bet if you or I with our big nonbird brains were to hide seeds in different places all summer and then find them again all winter, we'd do a lot worse than the scrub jays—probably fail to survive, if our lives depended on it. True, scrub jays don't have much else to do with their time. But that shouldn't take away from the fact that, like many other species, they have hyperdeveloped capacities that in us are weak or lacking altogether. It's no reflection on them that *our* best trick didn't just keep us alive, but lucked us into a different universe.

But because they have one mental capacity that goes way beyond us doesn't mean that their other mental capacities are equally developed, or even that they are developed at all. Too often people seem to think (or maybe feel emotionally, or at least behave as if) there's some universal fixed scale of intelligence on which all species can be placed somewhere, some higher than others, us higher than all. Evolution does not work that way. A repeated refrain in this book has been this: a species does what it has to do. If a bird enters the seed-caching niche it will probably get, sooner or later, the kind of skills a scrub jay has. Natural selection will see that it does. The ones who do what they have to do better than others will survive longer, breed more, have offspring that do even better than they did. The niche creates the intelligence—not some generalized cleverness, but whatever specialized intelligence the niche needs.

Well, birds are birds, you say; why would you expect to find human-like concepts there? You should be looking at our nearer relatives.

I would, if in their normal lives they did anything at all to suggest they had concepts. It's surely significant, if seldom noted, that the evi-

dence cited for animal concepts is seldom drawn from the great apes—
at least, not from great ape life in the wild. For that kind of evidence, we
have to go back to the monkeys.

Klaus Zuberbühler and his colleagues from the University of St.
Andrews in Scotland did a series of experiments that involved playing
recordings of alarm calls for leopards and eagles made by Diana monkeys (named after the goddess, not the princess) and the sounds made
by the predators themselves. I can't improve on Jim Hurford's summary,
so I'll just quote his words:

"On hearing first an eagle alarm call, then (after five minutes) the
shriek of an eagle, female monkeys showed less signs of alarm (giving
fewer repeat calls) than after hearing, for example, an eagle alarm call
followed by the growl of a leopard."

Hurford takes this to mean that Diana monkeys have concepts like
ours for eagles and for leopards. He reasons thus: for the monkeys to
behave differently depending on whether a predator sound is expected
(eagle) or unexpected (leopard) means that, for at least five minutes, the
animals must have kept the concept of "eagle" in their minds, and hence
they were shocked when what they heard didn't fit the threat they'd
been warned of.

Of course, that's one possible explanation. But there are others at
least equally likely. One, it's only surmise that the eagle and leopard calls
mean "eagle" and "leopard" to a monkey. They could just as easily mean
"threat from above" and "threat from the ground." The warning monkey
may be reacting not to concepts of eagle or leopard but to sounds that it
recognizes as threats coming from the air or the ground, respectively.

Two, receiving an eagle alarm call puts animals on high alert and
primes them to be ready with the appropriate strategy—if you don't
hide immediately, be prepared to dive into some bushes the moment
you see or hear anything above you. They will remain ready to execute
this strategy for some minutes after the warning, until enough time
passes without incident that they feel they're no longer in danger. A
predator sound that confirms the warning will simply keep them alert
and/or make them go for the bushes. It's the strategy that persists
through time, not the concept of a hovering eagle.

But suppose that instead of an eagle shriek they hear something they
weren't expecting: the sound of a terrestrial predator. This throws them

completely off balance, because the two escape strategies can both be fatal if used with the wrong predator. In the bushes, where you hide from eagles, leopards can grab you. Up a tree, where you'll be safer from a leopard, an eagle can easily spot you and pick you off. Small wonder Zuberbühler's monkeys showed less alarm when a signal was confirmed than when it was contradicted. They were alarmed in the second case because they simply didn't know which strategy to use.

Perhaps the best case for animal concepts like ours comes from the behavior of "language"-trained apes. You'll recall how, when they were first taught manual signs, it took them a long time to "get it"—several hundreds or even thousands of trials, spread over weeks or months, before they realized what the signs represented.

There are two possible explanations here. If people who think apes think like us are right, apes already had concepts of the right kind. They just didn't have labels for those concepts. Then kind humans came along and supplied labels. It took a while, but sooner or later came that "aha!" moment and the apes slapped the labels they were given on the concepts they already had—end of story.

The alternative goes like this. Apes didn't have concepts. Just like any other nonhuman animal they had categories into which they could sort things so they'd know how to respond to them. Those categories didn't jell into accessible concepts because they only functioned when the apes saw or heard or smelled or touched or tasted features on which the categories were based. That happened occasionally and unpredictably. The network of neurons that got activated when it did happen only linked up at those moments and quickly faded into oblivion when the features ceased to be perceived. Nothing remained that would tie all those features together.

Then the apes learned signs for their categories. The signs tied all the category features together and gave them a permanent home. They did so because the presentation of the category features—the features that distinguish, say, bananas from M&Ms—was no longer occasional or unpredictable. The researchers kept shoving bananas and M&Ms in the apes' faces. The neurons in the circuits activated by these presentations, together with those representing the objects' names, kept firing and firing. Neurons that fire together, wire together. The circuit was reinforced and anchored by the sign that had just been learned.

If that's all it takes to learn and use concepts, how come the trained apes didn't immediately start thinking like us?

To a very limited extent, they did. Thirty years ago, David Premack showed that "language"-trained apes could pass cognitive tests that untrained apes failed. But, if I'm right, it took us the best part of two million years to get from where we and the apes started—from no language at all—to where we are today. It's one thing to have representations of words/concepts stored in different parts of the brain. It's another for those representations to be linked by the afferent and efferent fibers that enable signals to pass in both directions from one to another, something that is vital if enough units are to be linked to make a coherent train of thought. Note that no ape has ever joined more than three signs in a communicative message. It's likely that they've never been able to merge more than three concepts into one coherent thought.

And there are dimensions beyond the cognitive that may be operative here. We won our language. Apes had theirs handed to them on a plate. We needed language to develop our niche. Apes didn't need it, never wanted it, need it now only to get rewarded and to keep their human keepers happy. And they've been using it for less than a human lifetime. All in all, I think they've done pretty well, don't you?

We should be able to respect them, without trying to turn them into blurred carbon copies of ourselves.

CONCEPTS AND THE DIVIDE

So let's assume that I'm right and that the presence or absence of human-type concepts is what divides humans from nonhumans. It's not really a question of problem-solving per se, but rather of the kinds of problems that are solved, the ways in which they are solved, and what kind of mental operations go into solving them.

For some reason, whenever we think about the evolution of intelligence, we tend to see it in terms of solving problems of ever-increasing complexity. We should be looking at the rarity with which animals solve problems by developing new behaviors, and the frequency—constancy would be a better word—with which we do it.

From time to time, of course, animals do produce new behaviors.

Japanese macaques wash potatoes (at least, some do). Foxes and coyotes kill pets in new subdivisions. Bears rifle garbage cans. Once upon a time in the West, Yvonne and I drove into an isolated campsite on a hilltop. It was high summer, so we were surprised to find it deserted. Then we saw the crushed and twisted garbage cans; some had been chained to steel stakes, but that hadn't stopped the grizzlies. We were out of there in seconds.

All these were new things, but all were accidental. For instance, bears can smell food a long way off, and American campers are wasteful. After the first few lucky finds, bears began systematically to target campgrounds. It's quite likely that stone tools started in a similar way. Protohumans bashed rocks, maybe in the course of nut-cracking (chimps do this regularly on the Ivory Coast), maybe as part of a dominance display (chimps do this with tree branches), maybe just for fun. Some of the rocks split and got sharp edges. Some smart ancestor realized you could crack bones with them, even cut stuff with the flakes from the core. So splitting and shaping rocks got to be a tradition, and for two million years paleopeople went on doing it. They got better at it, of course. Some of the pieces had slightly different shapes and sizes and gradually got modified, presumably for different purposes (although experts are often far from agreeing on what those purposes were). But there was a basic form common to all of them—they were all single, stand-alone pieces, longer than they were broad, with one end more (or less) pointed and one end more (or less) rounded. They were, fundamentally, variations on a single theme.

Now look at an Aterian point. Aterian points were made in North Africa starting perhaps as long as ninety thousand years ago. They're often described as arrowheads, but nowadays most people think it unlikely they were used on arrows, at least not the early ones. Probably at first they were spear points, then points on darts thrown with the aid of an atlatl, or spear-thrower, and only later became arrowheads.

At first sight, an Aterian point may look to you like no more than a downsized version of the old pear-shaped tools. But then you realize it's not a stand-alone piece. It's useless by itself. It has to be hafted onto a shaft of some kind, and that's new. You had to use maybe as many as four different types of material: stone for the point, wood for the shaft, mastic (a sticky resin from a bush that grows round the Mediterranean) and maybe gut or vine to bind point to shaft. You not only have to make the things you need; you can't make them unless you've figured out in advance how they could fit together and work together. You can't do that for the first time by trial and error, the way previous tools were first made. You have to work it out in your head—imagine it all, before you can start. To do that, you have to have concepts of the things you're working with and what you're going to do with them.

Look at the point more closely. Look at its tang. The tang is the part that fits into the shaft. Above it the point flares out with two flanges, almost barbs, before narrowing at the tip. Once the point pierces skin, that broadening will hold it there so the prey animal can't shake it loose. But the real function of the tang is to make a firm but narrow base that will fit into a space bored or split into the end of the shaft and filled with mastic so the tang fits inside it and can be bound for added security. The whole system, even before the atlatl, required forethought and planning. Forethought and planning in turn demand that you work not with physical objects but with your ideas of those objects—concepts you can move around in your mind to make new patterns and create marvelous and unprecedented things.

Now note precisely where the divide, the discontinuity, the boundary between human and nonhuman falls. Not between human ancestors and apes. It falls between our own species, on the one hand, and on the other, all other species that live or have ever lived, including our own immediate ancestors. Only our own species, it seems, has ever produced artifacts that needed forethought; therefore only our own species has ever practiced offline thinking.

Concepts versus categories

Critical here is the difference between a concept and a category. These words are often used very loosely, even treated as if they were interchangeable. In a moment I'm going to try to define them in neurological terms, because that's how we ought by now to be starting to define all those old-fashioned notions about things in the mind that we've been tossing around recklessly since before Plato.

For now, let's get a loose grip on them by merely saying that a concept is something you can "think about" and "think with," whereas with categories, all you can do is say whether something belongs in them or not. That's the difference. The similarity is that both terms refer to some kind of class into which things can be sorted—leopards, or tables, or grandmothers, anything at all. Because of that similarity, categories and concepts are sometimes treated as different names for the same thing. But if we don't distinguish between them, we'll never understand why humans differ from nonhumans.

Now look at all this through the lens of evolution. How can a brain best contribute to an animal's fitness? By telling it what's out there—what dangers it faces, what opportunities await its grasp. If the brain knows what's out there, it can tell its owner how to react. It's an X—eat it! It's a Y—up the nearest tree! It's a Z—freeze, and hope it goes away! Most of the time, of course, it's a W—no problem, go on doing whatever you're doing. But the brain's owner has to know. So it comes to divide things into classes—categories—that differ recognizably from one another (if it's an X, no way it can be a Y or a Z).

Let X be a squash and Y a leopard. Does the animal have two neat little packages in its head, one labeled "squash," the other "leopard"? Certainly not at first. In the early stages of brain evolution, the brain must have first picked up particular salient details: a kind of rapid movement, an unusual combination of colors. As senses sharpened and the ability grew to distinguish between things, even quite similar things, such details must have multiplied. Now a glimpse of a spotted coat through foliage, a distinctive cough, a particular swirl of movement in long grass, a pungent odor, the sound of paws landing on leaves as their owner sprang from a low branch—any of these or any combination of these could trigger the appropriate set of responses to an imminent leopard attack.

To be more precise, neurons in different regions of the brain, regions that dealt separately with sounds and sights and smells, would change their rate of firing in response to the incoming data, which in turn would trigger other neurons whose job it is to determine what the sensory neurons are talking about and what to do about it. And these decision-making units, if sufficiently excited, would then send, to neurons in the motor regions that control the animal's movement, signals that would indicate whatever response seemed most appropriate—freeze, flee, fight, climb a tree, or whatever.

Where's the concept of "leopard"?

You might say, "In the neurons that identify all the sounds, sights, smells, etc. as coming from a leopard." But do they, or is that just how we'd naturally think of it, since we're human and have a typically human kind of concept? Might it not equally be the case that the decision neurons are merely identifying "things on sensing which you'd better run up a tree"? And would there need to be any distinction between leopards and anything else that might make you want to run up a tree?

Let's be generous and allow that some neurons in the brain respond selectively to phenomena produced by leopards and only by leopards. Would they then represent a real equivalent of our concept of "leopard"—a concept that, if we choose, will link with every feature of leopards, their spots, their location, their hunting patterns, and on and on? Or would they represent only an identification—"It's a leopard!"?

Animals don't have to think about leopards once a particular leopard has gone away. They don't have to worry about what might happen when they next meet one, or devise elaborate plans for evading leopards. Remember how the vervet leopard warning means "leopard" only when there's a leopard there. Well, what I'm claiming here is that their communication directly reflects what goes on in their minds. It's not what Hurford and many other writers seem to think—that they have a rich mental life but have never found out how to communicate about that life. To the contrary, they can only communicate about the here and now because their minds can only operate in the here and now. They can't think, as we can, about leopards in the past or in the future or just in our own imagination ("I wonder if I could tame a leopard and have it as a pet?") because they don't have any sufficiently abstract mental units with which they could do so.

None of this means that nonhumans don't have a rich knowledge base, layer upon layer of memories, two or three different kinds of memory if it comes to that. If they didn't have such a stock to draw on, they wouldn't be able to function as well as they do. And nothing I have said should suggest that they don't have full access to these memories. Any of this knowledge base can be tapped, any memories triggered, by events in the world. A memory, once triggered, may trigger another, if that's relevant to the task at hand. What the animal can't do is think constructively about leopards when there's no real-life leopard around.

That's because there's no neuron or cohort of neurons that works as a pure symbol for "leopard."

In fact, the difference between human and nonhuman memory resembles in one respect the difference between RAM (random access memory) and CAM (content-addressable memory) in computers. In one, the user (read here the environment) supplies a memory address and RAM returns just the data stored at that address; in the other, CAM links that address with relevant data stored anywhere. (As you might expect, CAM is more complex and more expensive than RAM.)

So what exactly is there, in a leopard-identifying animal's brain?

I think there isn't anything in its brain that relates specifically to leopards in the way that either a thought or a word in the human brain does. All over the brain there are cohorts of neurons that respond directly to all the sights and sounds and smells that come in from the world by changing the rate at which they send out electrical impulses. Among all of these cohorts are neurons responding to sights and sounds and smells that might be made by leopards. When "enough" of these neurons ("enough" being still a black box) are triggered by a leopard appearance, the animal goes into high alert, may issue an alarm call, may take appropriate action. But the neurons activated on any given occasion are just one subset of the complete set of potentially leopard-responding neurons. The next appearance of a leopard may trigger a quite different subset, though the result (in terms of the animal's reactions) may be identical. Bottom line is, there's nowhere any fixed, determined set of linked neurons that represents "leopard" and nothing else.

But once you have a word or sign for "leopard," there has to be such a set—we'll see why in a moment. There has to be a fixed, permanent set of neurons that represent the sounds or gestures needed to produce the word or sign in question. But for that word or sign to have meaning, this fixed set has to link to all the different representations of leopard-bits on which the original "leopard" category was based.

In other words, I'm arguing that what started human-type concepts—things that have a permanent residence in the brain, instead of coming and going as and when they are stimulated—was the emergence of words.

Careful here. I'm not saying that "concepts are words," or "you have to have a word to have a concept." Least of all am I saying, "You can't think without words." Nonhumans do it all the time. They think online, run all kinds of computations on what they're doing. Imagine an eagle stooping on a running rabbit. While the eagle is in midfall, the rabbit changes course. The eagle has to recompute its trajectory in milliseconds. It may not be consciously aware of what it's doing, but if that isn't thinking, what is? You or I couldn't do it, that's for sure.

We too can think online; we can even think online and offline at the same time. Working on an assembly belt, driving a familiar route, we're on automatic pilot; we're thinking offline over things in our personal lives that don't have any connection with what our hands and feet are up to. The computations of time and speed and relative distance that we run while driving through traffic may be quite unconscious, though they don't have to be. The difference between online and offline thinking isn't unconscious versus conscious. The difference is that in online thinking, what's being thought about is right there in front of you, while in offline thinking it isn't.

Online thinking can be conscious or unconscious; when you're assembling a new piece of furniture from a list of printed instructions, it had better be conscious. But offline thinking has to be conscious, because by definition the things you're thinking about can't be there. Only the concepts can be there.

Maybe offline thinking *is* consciousness. But let's not get into that; we've got enough on our plate already. Let's get back to words. Words are reassuringly concrete, at least relatively so, compared with concepts and consciousness and suchlike, which tend to make you feel dizzy if you focus on them too long.

So all I'm saying is, without words we'd never have gotten into having concepts. Words are simply permanent anchors that most concepts have—a means of pulling together all the sights and sounds and smells, all the varied kinds of knowledge we have about what the concept refers to. But once the brain found the trick of making concepts, it no longer needed a word as the base for a new concept. It just needed some place where all the knowledge could come together and link with other concepts.

Once we had proper words (and I'm jumping the gun here; I still have to tell you how iconic "mammoth" sounds got to be words—I'll do that in the next chapter), here's what happened. The word had to have some kind of mental representation. There had to be a bunch of neurons somewhere that, when they fired, would start the motor sequence that would cause the vocal organs to utter "mammoth," or whatever. And that bunch of neurons had to be permanently accessible, had to be willing and able to fire whenever they were asked to do so.

SUMMING UP

I'd be the last person to pretend to you that the issues we've discussed are cut and dried, or that the answers to the questions I've raised here are plain and straightforward. In order to make the points I needed to make, I've had to simplify many complex things. In order to save this chapter from bogging down in a morass of detail, I've had to downplay or ignore topics that many experts in the field concerned will regard as of paramount importance. I still think I took the right course—the only course, if we're to see the woods and not just the trees, if we're ever to get a grip on what makes us so different from other species.

The only test of a story is its explanatory power. The best story is the story that explains the most things, that passes the greatest number of tests for what an explanation should accomplish. Before we look in more detail at how language and thought coevolved, I want to summarize where we're at and give a few compass bearings for where we still have to go.

The main point to be borne in mind is that between humans and nonhumans there are two discontinuities, not just one. We have language and no other species does, and we have seemingly limitless creativity and no other species does. Language and creativity are both, for all practical purposes, infinite; is this mere coincidence? For two independent discontinuities of such size to exist in a single species is something altogether too bizarre in evolutionary terms. So at the very least it's worth exploring the possibility that the two discontinuities spring from the same source.

Language involves the mind and creativity involves the mind—the mind being no more than the brain at work. So the likeliest cause of such a double discontinuity would seem to lie in a difference between the workings of human and nonhuman brains. One possible difference, one that would seem to give rise to all the phenomena we've been look-ing at, is that nonhumans have categories and humans have concepts.

Categories sort things into classes but can only be evoked by physi-cal evidence that members of those classes are present.

Concepts sort things into classes but can in addition be evoked by other concepts even in the absence of members of any of the classes con-cerned. Hence they become available for offline thinking.

All the things nonhumans do that make it look as if they had con-cepts like ours can be explained by feats of memory, specialized and dedicated mechanisms for solving problems posed by niches, stereotyp-ical strategies responding to different threats, and/or other causes or combinations of causes that at no point entail the possession of con-cepts.

Eventually, language and human cognition did coevolve. But first, the first words had to trigger the first concepts and the brain had to pro-vide those concepts with permanent neural addresses. Only then could the creation of concepts enable the mind to roam freely over past and future, the real and the imaginary, just as we can do nowadays in our talking and writing. In other words, before typically human ways of thinking could grow, language itself had to grow. And in the next chap-ter, we'll see how.

II

AN ACORN GROWS
TO A SAPLING

THE TRIPLE UNCOUPLING

At the end of chapter 8, I asked how such a small change in the way protohumans communicated—the tiny handful of signals required by recruitment—could have developed into anything as complex as language is today.

I can answer that question in just four words.

With the greatest difficulty.

If you believe that animals have minds with concepts just like ours, it should have been easy. Most people assume, as I did before I really thought about it, that once you realized what a linguistic symbol was, everything would be simple and straightforward. As soon as some kind of protolanguage got started, it would take off. All that would be involved was a slapping of linguistic labels on an array of concepts that were sitting there waiting for them.

After all, hadn't apes, once they got over their initial bafflement, proceeded to pick up signs on a minimal exposure? If they could do this with brains less than a third the size of ours, why couldn't our ancestors, with brains up to twice as big as those of apes, have done the same?

The adaptive benefits were beyond question. All the things I've dismissed as possible selective pressures for *actually starting* language—instructing the young, competing socially, displaying sexually, making artifacts, gossiping, performing rituals, and so on—were things you could *use language for*, once you had it. All these activities would be enhanced, some a little, some immeasurably, by a species whose discourse could range back and forth in time, hither and yon in space. Surely one after another, each of these activities would acquire its appropriate words. And while the absence of any kind of regular structure might keep actual utterances down to a few words at a time, you should soon have quite a respectable protolanguage.

I cannot now see a single compelling reason for believing this, and there is much that points in a very different direction. The most impressive evidence for a long, slow gestation of language can be found in the most tangible data we have—the fossil and archaeological record of our ancestors over the past two million years.

The long stagnation

When paleontologists describe those two million years in books aimed at a popular audience, here's the kind of picture we usually get:

> Long-continued increase in size and complexity of the brain was paralleled for probably a couple of million years by long-continued elaboration and "complexification" . . . of the culture. The feedback relationship between the two sets of events is as indubitable as it was prolonged in time [Phillip Tobias, South African paleontologist].

> The fossil and archaeological record for *Homo* picks up around two million years ago in East Africa. And what a record it is! Brain size "took off" and subsequently doubled from approximately 700 cubic centimeters to 1400 cubic centimeters . . . Recorded tool production also accelerated in *Homo*, spanning from initial clunky stone tools to contemporary computer, space and biological engineering [Dean Falk, professor and chair of anthropology at Florida State University].

Tell that to a Martian, and he'd probably assume that things like stone bridges were invented about a million years ago (it *was* the Stone Age, wasn't it?), the wheel (*Flintstones*-style) maybe half a million years ago, and steam trains about a hundred thousand years ago. I've never understood why reputable experts in the field say things that they know perfectly well are not true. Is it because the facts about human evolution are so totally at variance with anything you'd expect that paleontologists simply don't want to admit them? I can't think of any other explanation.

Tobias is right that the increase in brain size went on for the best part of two million years, but entirely wrong when he claims that "the culture" elaborated and complexified. Falk is right that brain size doubled—actually, more than doubled, in Neanderthals—but misleadingly suggests that the progress from "clunky stone tools" to modern technology was smooth and evenly paced. It wasn't.

As I mentioned in chapter 7, the standard tool of *Homo erectus* was the symmetrical, pear-shaped object known as the Acheulean hand ax. For more than a million years, this remained unchanged. And the few other tools that developed, the so-called borers and scrapers, were basically variations on this tool, making it thicker or thinner, more pointed or more long-edged. There were no notched or tanged points, like the Aterian points described in chapter 10. There were no hafted tools—tools that required two or more parts to be joined. There were no artifacts made from bone or ivory—nothing but variations on the one-stone tool. During the entire period, about the only innovations were the taming of fire, the invention of spears, the erection of primitive shelters, and the start of serious big-game hunting. Even for these milestones, firm evidence comes only from the latter part of the period.

Yet more than a million years ago there were already plenty of hominids whose brain sizes, more than 1,000 cubic centimeters, fell within the normal range of modern humans. If increased brain size spells increased intelligence, and if people with brains smaller than those of some members of *Homo erectus* can talk and write and do science and invent stuff, how is it that the species that preceded ours endured the same harsh conditions of life, without any serious attempt to improve its lot, for the best part of two million years? For far from the smooth progression Falk claims, everything in human civilization—herding, agriculture, cities, industrialization, and the exploration of the solar system—was squished into a bare two-hundredth part, 0.005 percent, of that period.

If language is indeed what drives human thought, and if language began two million years ago, how can such things be?

Well, you could argue that I'm wrong, that language did not start two million years ago, that it did not start until much later. But then we're back with the problem, what else could have started it? I know of no development in prehistory later than power scavenging that could have served as a trigger. Of course that doesn't mean that there wasn't one. But there isn't even a promising candidate in sight. Unless one turns up, power scavenging remains the likeliest cause.

Besides, the later the birth of language, the harder it becomes to find time for the extensive rewiring of the brain that language required. Short of some magic mutation, that process must have taken quite a while. Moreover, as we'll see in subsequent sections of this chapter, there were hurdles over and above brain rewiring that would have delayed things further. It's tempting to hypothesize a rapid and relatively recent coevolution of language and culture, starting from scratch, over the last hundred thousand years or so. Such a scenario would feel comfortable, would fit well with contemporary thinking. But if we ask, "What evidence supports it?" the answer is "None."

There is still another alternative. Some think that language developed fully at a much earlier date. They will point to some of the hunter-gatherers who still survive in the modern world. In language and intelligence, these are completely normal modern humans. Yet their tool kits and all the things in their material culture are hardly more complex than those of Cro-Magnons. So why shouldn't our remote ancestors have sat around the campfire hundred of thousands if not millions of years ago, with a modern language capacity or something not too far short of it, telling stories, politicking, wooing one another with words, happy as Larry in their free hunter-gatherer lives, simply preferring not to weigh themselves down with all the massive impedimenta of civilization?

Because hunter-gatherers lie at the conservative end of a spectrum of behaviors—a tiny minority who, by chance or by choice, failed to innovate in the way the vast majority of our species did. Previous hominid species did not show any spectrum of behaviors. Their behaviors varied as little as do those of chimps or vervets or any other nonhuman species, whether over eons of time or thousands of miles of space. The idea that a species could have a highly adaptive capacity that none of its members ever exercised goes against all we know about how species construct their niches. To the contrary, species exploit their capacities to the full, even expand on those they already have. Any species pushes the envelope to the extent that its genes and the phenotype they produce allow it. It's unthinkable that any species should have the power to radically alter its behavior but that no member of that species would ever use that power. So the idea of a linguistic species with next to no technology is as problematic as that of a species developing language from scratch at a very recent date.

So, on balance, I'm convinced that an early start coupled with a painfully slow development of language fits better once all the evidence we have today is taken into account.

Barriers to language

To understand why that start was so slow and difficult, we should look a little more closely at what the recruitment procedure actually provided. It was not a sapling, not even a young shoot. It was more like an acorn—something that, with good luck and nourishment, might one day expand into a tree. But the future shape and form of language, or for that matter even of protolanguage, was no more visible in the handful of recruitment signals than the shape of the oak tree it will one day become is visible in an acorn. Recruitment had broken the mold of the ACS; that was the crucial step. And it had broken that mold not in a bee or an ant or any species with a microscopic brain, but in the one that currently had the highest ratio of brain to body of all the species on earth.

But that species wasn't smart enough to know what had happened to it. Its members, unlike the trained apes, were not surrounded by another species, one that already had language, one that was single-mindedly determined to teach language to them. They were pioneers. They were alone in the universe. They couldn't have had the faintest inkling of the possibilities their discovery had opened up for them, and there was no one there to take them by the hand and show them those possibilities. They were probably not even aware that they'd done anything new.

So let's look more closely at recruitment signals, see what they had and could do and what they didn't have and couldn't do. The signals

- had "functional reference" in that they specified one or more—probably several—megafauna species whose exploitation required recruitment. (Functional reference had, you'll recall, already been achieved by the specific warning signals of vervets and other primates, although not by any of the great apes.)

- had displacement, something that hadn't been achieved by any primate species; they contained information obtained well ahead of utterance, about things well outside the recipients' sensory range.
- were created and learned, rather than hardwired.
- contained protonouns—names of species—and probably also protoverbs: noises and gestures that could be interpreted as "Come!" or "Hurry!"

That's the plus side. However, in the first chapter I pointed out three further properties that had to be achieved before even a protolanguage could begin. Signs would have to be uncoupled from situations, from current occurrence, and from fitness. In fact, signals were still

- coupled with situations; certainly at first, perhaps for a long time, they were used only for recruitment, and thus made sense only when there was a large dead herbivore not too far away and other group members had found it and were trying to get attention.
- coupled with current occurrence; even though displacement had been achieved, you couldn't yet talk about the large dead herbivore you found last month, or suggest strategies for finding more such in the future. You couldn't talk about such things until you had words of some kind showing that you *were* talking about the future or the past.
- coupled with fitness; even though power scavenging involved cooperating with non-kin, it still contributed to individual fitness, precisely because, without cooperation, each individual would be deprived of possible food, and with it, each individual would benefit.

Breaking those couplings was an essential prerequisite for the development of even the simplest kind of protolanguage. But as we saw in the previous chapter, the couplings weren't there by accident, each with its own connections that could be easily snapped. They all stemmed from one cause—the inability of prehuman minds to deal with anything other than the animal's immediate circumstances. In other words, they were incapable of displacement.

Let me dwell on this a while, because it took me a long time to fully grasp it and I suspect I was in good, perhaps almost universal, company.

When most people look at language and contrast it with ACSs, displacement is not often picked as its most salient feature. People think of learning; ACSs are innate, language has to be learned. People think of arbitrariness; ACS signals often show a direct relationship to what they signify (cringing postures to indicate submission; loudness, repetitiveness, or intensity of vocal calls to indicate firmness of purpose). In language, however, words of similar meaning—dog, *chien*, *perro*, *Hund*—bear no obvious relationship to what they describe or even to one another. People think of combinability; ACS signals won't combine at all, while words, phrases, and clauses will combine without limit. People think of complexity; while language has an intricately layered structure of sounds and units of meaning and syntax, ACSs are single-level, what you see is what you get. Displacement gets second billing, if it gets any billing at all.

It's only when you fully appreciate what displacement means, how the absence of displacement is not just a casual feature of ACSs but a crucial defining feature of prehuman minds, that you can start getting the complete picture. That picture shows you two complementary things. It shows you how achieving even the rather superficial form of displacement found in recruitment signals was the greatest single step any communication system could have made in the direction of language. But it also shows you the immense difficulty, even when that step had been taken, that's involved in creating true displacement, true escape from the here and now in which all species had hitherto been trapped. To do that, you had first to make concepts, mental symbols of reference no longer bound by particular instantiations of the things referred to. Only with such abstract symbols could you roam mentally, freely through space and time as we do today, in both language and thought.

Still, the displacement in recruitment signals formed a wedge already driven deep into the status quo—a wedge without which we would either still be wandering houseless across the savannas, or more likely would long ago have gone extinct.

Let's see how that wedge could have worked.

FROM SIGNAL TO WORD

In the initial, recruitment phase of protolanguage, there were, properly speaking, neither concepts nor words. Recruitment signals weren't words. They were iconic and/or indexical signals that, to those who used them, were no different from all the other ACS signals they already had. Signals had to become words and words had to give birth to concepts before anything you could even call a protolanguage could be born.

The signals associated with recruitment were the only signals in the protohuman ACS that had displacement, and in the beginning they were tied to what had happened or was about to happen. What you might loosely want to call the "mammoth" signal might have been better interpreted as "We've just found a dead mammoth and we want you to come help us butcher it." But precisely because the signal— pantomime of the live animal's looks or movements, vocal imitation of the noise it made, or whatever—drew attention to the nature of the animal rather than simply pointing to an actual appearance of that animal, it became available for use in other circumstances that had to do with mammoths.

There's no recourse here but to tell just-so stories. Walking along a dried but still muddy watercourse, brief fruit of rare rains, an older and a younger individual see a set of deep footprints. The older one points and gives the mammoth signal.

A bunch of young ones, some little more than infants, are playing. One or two are old enough to have gone out on megafauna butchery expeditions. (How old did that have to be? My guess is, they needed all hands, so as soon as you could run fast enough and throw far enough, you were on.) The older ones mime the expedition, interspersing their mammoth signal with boastful gestures and noises. The younger ones listen and imitate; in their still-plastic brains, sounds morph into images and out again.

A group comes upon a heap of big, stripped bones. This was one they didn't find in time. Some of them, in tones of disappointment or anger, make the mammoth sound as they turn the bones over, looking for scraps the other scavengers may have left.

Gradually the sound is getting divorced from the situation that gave rise to it. Put something into enough different contexts and its particular details become blurred; it gets closer to becoming an arbitrary symbol.

At the same time, a representation is formed in the brain: a representation of the mammoth sound. How is this different from the representation of all the other ACS signals? At first, not at all—except that it's been learned; it's passed from generation to generation. For the moment, that doesn't carry any consequences. If there are other signals that go with it—signals that might mean "Hurry!" or "Come!"—they go with it only in the recruitment scenario, or in the pantomime reenactment of it (which might have been done, as ritual, by adults as well as children, but we've no idea when rituals began). However, as the intimate connection between signals and situations begins to erode, those signals become increasingly more wordlike, and increasingly available to the combinatorial process that is essential, even in protolanguage.

But what we must bear constantly in mind is that we're not dealing with a species that was sitting around passively waiting for happenstance or genetic drift to carry it along. We're dealing with a species that was actively carving out the niche of high-end scavenging, and this process in turn fed into the growth of protolanguage.

NICHE CONSTRUCTION DRIVES LANGUAGE

Think about it. Here's a species that's developing a new niche. Alone among species, it can drive other scavengers away from dead megafauna long enough to access the carcass and make off with the best cuts. This species has the biggest brain going and the carcasses form its richest possible supply of food. So this species is going to go on, millennium after millennium, scanning the skies for cruising vultures but otherwise just accessing whatever carcasses it happens to stumble upon. Whoops! Look at that! Why, I do believe it's a dead giant hippo!

I don't think so.

I think, before very long, it occurred to some brighter-than-average *erectus* to take a more proactive approach. After all, wherever the megafauna wandered, they left a pretty big signature—dung heaps, trampled grass, torn-up vegetation, footprints around streams and rivers. Some maybe had regular migration patterns. The logical thing to do was not to wait until you stumbled on a dead one, but to track the herds, pick out sick or injured ones in advance, be there when the dying commenced, and get to work probably before the moment of death, as soon as the animal became unable to defend itself.

We do know, and it's one of the better established facts in archaeology, that it was in the period between 2 million and 1.6 million years ago that three things coincided:

- the appearance of the Acheulean hand ax
- the appearance of tooth marks superimposed on cut marks
- the switch from catchment scavenging (the intensive exploitation of relatively small territories) to territory scavenging (ranging over much wider areas)

The hand ax was the all-purpose butchering tool, used to hack through bone and sinew once the hide had been cut open by flakes, and also as a projectile to drive off the scavenging competition. The cut marks superimposed by tooth marks are the unarguable signature of high-end scavenging—irrefutable evidence that human ancestors had gotten thar fust with the most. The switch to territorial scavenging could have been made possible only by a switch in resource exploitation— meat taking the place of bone marrow. Although it's hard, and probably always will be, to find smoking guns, direct evidence that power scavenging became our ancestors' main activity, the balance of what evidence there is points in that direction.

For a species that had become dependent on scavenging large carcasses, it would become more and more important to accurately read all the signs that the megafauna left, to determine species identity and relative age, numbers in the group, things that indicated an animal might be injured or sick. Disputes would inevitably arise about how the signs should be interpreted. How old were they? How many animals would have made them? Should we follow group A, small but with one animal in it that might be sickening, or group B, much larger but with all members in apparent health?

Such disputes would intensify once a potential victim had been identified.

The subgroups you'd have to recruit would have their own agendas. They too might have a sick animal in view. The task of persuading a subgroup to drop what it was doing and come join you would get harder, not easier. How much nearer death is your animal than their animal?

Then there was the competition. Were there other potential scavengers around? If so, how many? And perhaps even more crucial, what species did they belong to? Here again, iconic signals go a long way, but precisely since they are iconic, not indexical—not, that is, pointing directly at a particular exemplar of something—they're easier to convert into symbols as they get to be used in more and more contexts. If you can name major predators and indicate the signs they leave, you can teach your children lessons that may in the future save their lives.

In this way, developing the high-end scavenging niche would have both created new words and deployed old words in new contexts, further weakening the uncoupling of words from situations, from current occurrence—even from fitness.

COMPLETING THE TRIPLE UNCOUPLING

This couldn't have been a rapid process. The coupling with fitness would have been perhaps the longest to break. Imparting information purely for information's sake still lay ahead.

The French scholar Jean-Louis Dessalles has proposed one factor that would have encouraged such a process. Primates are highly competitive, constantly seeking to be one up. Modern humans can achieve one-upmanship by being the first to pass on some piece of new and significant information. Surely, Dessalles argues, this behavior should go back as far as there were ways to give information.

As a starter for language, this proposal seems at first to fall foul of the condition that fails so many promising explanations of how language began. There's little doubt that the behavior would have helped to expand language at later stages. However, until language had been up and running quite a while, there simply wouldn't have been enough words for the proposal to work.

But it does provide a likely source for the *creation* of new words once protolanguage had gotten started. Relatively rare yet recurring events that severely impacted a group—flash floods, hurricanes, grass fires—were things whose effects could be mitigated by advance warnings. Moreover, ability to give advance warnings would confer prestige on the giver. Consider flash floods, for example; any sound or gesture that reminded the group that a sudden, heavy fall of rain likely preceded an inundation would have been stored and repeated.

It's still about fitness, you say. But suppose that while a group is working on a megacarcass, there's a sharp shower, and an immature, rather nervous hominid gives the flood call. Its fellows begin to head for higher ground. Quickly, wiser elders couple the flood word with a dismissive gesture, one perhaps that already means "Don't do that!" So negation is born. "There isn't going to be any flood, this time" is still a long way ahead, but that's what's meant. And a multiplication of words, plus the power to combine them, is all that is lacking now.

A just-so story, of course. But negation has to come in somewhere. It's one of the first things small children learn—develop would be a better word, because they don't negate like mother does, with "don't"s; they just stick the general negative word "no" in front of whatever it is they don't want. In the scenario I described, our older and wiser hominids would be using the flood word not to refer to an actual or imminent flood but to assert the current nonexistence and future unlikelihood of a flood. And nonexistence is something no ACS can handle. It can't handle it because ACSs refer, if they refer at all, to things that actually *exist*, in the here and now—"mind-independent entities" that have a physical life in the real world. But "no flood" can't refer to any flood there ever actually was—only to the abstract concept of floods.

These are just some of the steps that might have been taken to divorce words from any real-world things they referred to and make them the outward forms of true concepts, applicable to any or all members of a given class, whether hypothetical or real. Until the triple uncoupling was completed, all you would have would be just a more sophisticated version of bee "language" or ant "language"—a system dedicated to one major purpose (providing food for the group) but useless outside of that function.

I didn't always think this way. I supposed that if you gave the equivalent of bee language to a big-brained animal, that animal would rapidly expand said language to help out all the other functions big-brained animals perform. Somehow that seems very natural to members of our species, a species whose members are saturated in language from a time before our memories begin. But what basis do we have for such a notion?

Language was an unforeseeable development whose properties ran counter to any behaviors that had happened before. It was an evolutionary anomaly at least as great as the emergence, in a world filled exclusively with single-celled creatures, of the very first multicellular organisms. Greater, in fact; multicellular organisms did no more than multiply what had been there before. Language, on the other hand, was a pure novelty.

I now suspect that what recruitment brought in its train—a system purpose built, devoted solely to the exigencies of power scavenging—may have persisted for several hundred thousand years or more, while hardly yet deserving the name of protolanguage. A kind of hybrid stage, halfway between ACS and protolanguage—little more developed than the "languages" of bees or ants.

PIDGINS FLY TO THE RESCUE FOR REAL, THIS TIME

Before this occurred to me, when I had to explain the long stagnation while still assuming that a relatively rich protolanguage developed relatively early, I naturally had recourse to Dan Dennett's indispensable invention, figment (no writer on human evolution should ever be without it, or ever *is* without it).

My particular piece of figment went like this: without syntax, you can't put thoughts together any more than you can put words together. In order to create new cultural and technological stuff, you first have to put thoughts together in an orderly and disciplined fashion. In order to do this, you need syntax. Alas, our poor ancestors, without syntax, could not put their thoughts together. Therefore, they were unable to think the kind of thoughts that they needed to think in order to achieve real cultural and technological innovations.

Thence sprang the barbed-weapon scenario, designed to show why, without syntax, we could never have invented barbed weapons. I used it in several talks, and it went something like this:

Any barbed weapon (dart, arrow, fishhook, harpoon) dates to within the last hundred thousand years of prehistory—in other words is indubitably the work of modern humans. What would you have to think, in order to even get the idea of making a barbed weapon? Surely, early *Homo sapiens* thought something like this: "When I put a smooth point into an animal, often the animal shakes itself and the point falls out. If the point falls out, the wound closes and no longer bleeds, so that the animal isn't weakened and may escape. If I made a point in such a way that it would stay in the prey, then it couldn't be jerked out, the animal would go on bleeding, be weakened, and either fall down or be captured. In the long grass I just walked through, burrs clung to my legs, seeds that have attached to them a little thing that catches in my skin and doesn't fall out. What a great idea to make points with something like that!"

When I used this example in talks, nobody ever stood up to state the obvious: What was to stop anyone thinking something like the following?

"Me throw dart/spear. Point hit animal. Point fall out. Wound close. Animal get away. Suppose point stay. Animal bleed. Animal get weak. Catch animal. Look this seed. Seed stick in skin. Seed get little thing. Little thing stick in. Suppose point get same-kind thing. Maybe point no fall out. Me catch animal. Me kill animal. Me eat animal."

If what you just read sounds like some kind of pidgin, fine. Although for a variety of reasons it's far from a perfect model, a pidgin is the nearest thing to protolanguage that we're going to find in the modern world. Unfortunately, good pidgins are hard to come by in the modern world. I'm sure that everywhere there are incipient pidgins, wherever people with different languages come into contact with one another, but these seldom if ever develop into the full-blown thing nowadays. English is killing pidgins maybe even quicker than it's killing established languages. Why try to start a new language if there's a ready-made one already spreading across the globe like crabgrass?

I was fortunate enough to arrive in Hawaii before the last of the good old-time pidgins disappeared, and I was sometimes amazed by what it's possible to do with quite limited resources. Here's just one example, a philosophic meditation on the vicissitudes of life using virtually no syntax and only twenty-two different words: "Some time good road get, some time all same bend get, angle get, no? Any kind same, all same human life, all same—good road get, angle get, mountain get, no? All, any kind, storm get, nice day get—all same, anybody, me all same, small-time."

Granted, this was a modern human fluent in his original native language (Japanese, as it happens) and granted too that you wouldn't expect such flights from *ergaster*. I merely wanted to demonstrate that you can do far more with a tiny vocabulary, just stringing words together, than you might at first have thought.

Just what do we gain by putting this into modern syntactic form?

"Sometimes [when you're traveling], you drive on good roads while at other times you meet with obstacles like bends and sharp corners, don't you? Everything else is like that, human life is just like that—sometimes you find good roads, other times you find things that are like corners or mountains, you experience different kinds of weather, sometimes there are fine days and sometimes storms, aren't there? Well, life's just like that for everyone, same as it was for me when I was young."

It's longer, it's more verbose, prices we pay for eliminating obscurities and ambiguities and giving a smoother flow to things. In terms of the actual sequence of ideas, what's to choose between the two versions? The first may lack grammatical structure, but it has as much semantic and pragmatic structure as the second, and which of those factors would be most important in thinking? Maybe thinking without syntax is less fluent than thinking with it. But, given the right words and enough of them, couldn't the predecessors of modern humans have done just a little bit better than a million years with the same old hand ax? Wouldn't they have done *something* to break what one paleoanthropologist referred to as "the almost unimaginable monotony" of the Lower Paleolithic, the Old Stone Age?

The conclusion to which this seems to lead us is that protolanguage might not have reached even the level of an early-stage pidgin until our own species emerged. Or it may simply mean that, for reasons still unclear, the ability to connect words, to construct short and practical messages, emerged long before it became possible to link concepts into coherent trains of thought. That's just one more of the many questions that we cannot yet answer.

Things you can do with words

As I said in chapter 2, I eventually came to agree with Terrence Deacon that it was symbolism rather than syntax that marked the boundary between humans and nonhumans. Originally I'd been skeptical about Terry's claims; it seemed to me that the one thing apes couldn't handle, and humans could, was syntax. But suppose it was the case that there was no hope of getting to syntax until you had symbols—and not just symbols that you'd had handed to you on a plate, the way chimps got them, but symbols you'd had to fight for and win over hundreds of thousands of years of glacially slow progress? Once that thought—what you might call the "can't get there from here" hypothesis—had entered my mind, it began to seem foolish to get down on the chimps for failing at something they could never have been expected to do. Moreover—and this is one of the tests of a good thought—it started me thinking in new and productive ways about what it would really have taken to get from no language even to protolanguage.

What words would have come first in protolanguage, and would it have made any difference?

Remember all the things that have been advanced as possible selective pressures for language—child care, toolmaking, gossip, hunting, social maneuvering, and any more you can think of. Now even if, as I've proposed here, none of these did in fact select for language, it remains the case that they were all matters of greater or lesser concern for our ancestors, and they were all areas where, if you already had the beginnings of language, more of it would make a difference.

Now do this thought experiment: make lists of what would seem to you to be the ten most useful words in each of these and other similar areas of interest. Put your lists side by side and see how much overlap there is.

My prediction—very little, maybe even none. The protolanguage would have to be at least as big as the sum total of these lists. But even then would it have been big enough to be useful?

To find out, ask yourself how much interesting gossip or useful hunting information you could convey even with all of those ten words, or to what extent you could use them to enhance your status in your social group. Do it the brute force way, just like I would: make all the theoretically possible combinations of these words and see how many of them would actually make sense. (Don't cheat now—no syntax, and no "if"s, "and"s, and "for"s or any other of those meaningless words that glue the meaningful ones together and allow us to interpret them swiftly, quite automatically, and—most of the time—unambiguously.)

Then ask what might have motivated the creation of any one of these words. "They wanted to talk about X" won't cut it. They couldn't yet know what "talking about X" meant, let alone want to do it. You have to try to think of some practical problem the new word might have solved, or some situation that might have pushed someone to create a particular word—anything at all that might have driven the word-making process.

Believe me, you'll be doing research that nobody's done yet. You see, it's far easier to talk in a vague general way about how protolanguage would grow and how it might work than to get down to the nitty-gritty, the sheer nuts-and-bolts of how all the plausible-sounding things people are constantly saying about early language might work out in practice. You'd think that people who confidently say language resulted from this pressure or that pressure would at least think about what words the pressure in question would have had to evoke, and if it's reasonable to suppose that naive hominids not that far removed from other apes could have or would have invented those particular words. But until this book, nobody, literally nobody, who ever wrote about language has ever suggested what the first words might actually have been, or under what precise circumstances they might have been uttered, least of all how those words might have fitted the particular selective pressure that the author, with such confidence, put forward as the crucial engine for language evolution.

My guess is that none of our ancestors were actually trying to build a vocabulary, because none of them could have known what they were doing. While recruitment signals might in principle have triggered an "aha!" moment, a realization that now symbols could be linked with things other than big dead herbivores, there's no compelling reason to believe that this actually happened. To the contrary, when we consider that the hominids in question were a lot closer, in minds and behavior, to apes than to us, it seems hardly likely that anything could have happened in quite so quick and dramatic a fashion. And when we recall what we saw in the previous chapter, that there's good reason to believe that words did not follow but preceded concepts, the chances of an "aha!" moment sink close to zero.

What is needed here is the opening of a new field of inquiry—the study of hypothetical early vocabularies and the communicative consequences of different word choices in forming vocabularies of a hundred words on down.

The beauty of this field is that you can actually do experiments: you can give such vocabularies to real-life subjects—not just simulated cyber-agents, but flesh and blood people—and get them to perform varying types of communicative tasks with these limited resources. Of course you can't actually replicate Stone Age languages—we're humans and they weren't—but you'll at least establish upper limits on what could have been produced, and I'm willing to bet that in the process, facts we never knew about language will emerge.

The only person I know of who's doing this kind of research is Jill Bowie, who's just finished her doctoral dissertation at the University of Reading, England. So far, she has presented people with a fifty-word vocabulary and a *Survivor*-type scenario, and required them to communicate using only that vocabulary. Here's just one example: one of her subjects, needing to express "In the forest, Fred was bitten by an enormous snake," produced "Many many tree (CIRCLES HANDS IN AIR TWICE)/Fred/snake/big snake/big big snake/(TAPS LEG)." The mix of modalities is particularly impressive, rendering moot all those arguments about whether language was originally signed or spoken or mimed. Answer: all of the above.

More such experiments should be carried out, varying both the content of the vocabulary and its size. Which words, and how many of them, would be needed for significant communication in the different areas of human activity—child care, social intrigue, toolmaking, exchanging gossip? It would be possible to test empirically what I've claimed here, that none of these activities could have triggered language, either because words required by them are too abstract to be plausible as early inventions, or because too many words would have had to be invented before any useful or interesting messages could be exchanged. Since those same activities, though highly implausible as prime causes, could still have contributed to the expansion of protolanguage once it was up and running, the likely kind and extent of their contribution might also be assessed.

Possible pitfalls abound, of course. One species, no matter how smart, can never think its way back into the skin of another. Facile interpretations, slick just-so stories—these and more will bedevil the inquiry. The belief that sheer contingency rules wherever humans are concerned, making reconstruction futile, threatens the work from an opposite direction, and must also be resisted. That may be a valid belief, but we'll never know if we haven't tested it. And if we can unravel what happened in the first few seconds of the universe, discovering what happened at the beginning of protolanguage should not lie altogether beyond human powers.

CONNECTION AGAIN

But what the foregoing assumes is that from a very early stage words could be connected with one another to form more complex messages. Is that a reasonable assumption? Granted that apes and young children learn how to do this without explicit instruction, we must still bear in mind that both have numberless models—parents and trainers—to show them that it's possible. And as we have seen, no ACS can properly combine any of its units. Aren't we assuming too much here?

I don't think so. The reason ACS units can't combine does not arise from any mysterious constraint on combinability. It arises from two simple facts: ACS signs are complete in themselves, and combining them would make no kind of sense.

For words, these conditions are reversed. Words are seldom complete in themselves, and in isolation may mean a variety of things. A man driving on a country road met a female driver coming in the opposite direction, and she shouted something at him, of which he heard only the word "Pig!" Not unnaturally, he thought he was dealing with some rabid feminist obsessed by male chauvinism, and was forced to brake violently when, around the next bend, he saw an enormous porker lying in the middle of the road. While imperatives like "Stop!" or "Run!" may be unambiguous, most words acquire precise meaning only when coupled with other words.

Animals can combine actions into a series, so why not words, once they can acquire them? Since the simplest way to combine things is to string them together (a process Chomsky dismissed as logically impossible, remember?) we can safely assume that that was how protolanguage did it. It's almost certainly the way apes and pidgin speakers do it. And if I'm right in what I conjectured in chapter 10—that concepts not only have to be established but must be neurally linked to one another before serious thought or language can appear—it's necessarily the case that apes and hominids should do it that way. (Pidgin speakers, of course, resort to beads-on-a-string linkage not for this reason but because pidgin, unlike their native languages, has no automated system for creating hierarchical structures.) So protolanguage, after perhaps a million years or more, would have looked something like a pidgin, with words not significantly different in meaning from modern words.

Of course they might not have sounded at all like modern words. Modern words form large vocabularies and have complex sound structures. These facts are interconnected. The more words you have, the harder it becomes to distinguish one from another. Consequently, a trade-off is involved between speech sounds and word length: the fewer the sounds your language uses, the longer the words must be if they're to be distinguished from one another.

Modern languages make this trade-off in different ways. They have a range of distinctive speech sounds that stretches from 11 (Rotokas, spoken in Papua New Guinea) to 112 (!xoo, spoken in Botswana). But note that what actually happens in given languages provides inadequate clues to the full biological capacity of humans. A Rotokan baby, placed in infancy among the !xoo, would surely grow up speaking fluent !xoo. So the ability to make and distinguish a wide range of speech sounds now forms part of human biology—we could all of us have potentially made and distinguished all of the 112 sounds of Ixoo, even if at our present age that's become an impossible feat.

In other words, any increase in vocabulary would have selected strongly for an increase in phonological complexity. And this, in the later years of protolanguage, would have begun one of the processes that would eventually distinguish true language: the double layering of sounds and words.

I'm imagining that the earliest protolanguage words—as distinct from the manual signs and other signals of meaning—would have been indivisible chunks of sound, sharing no features with other words. If this condition held, there couldn't be very many words. Beyond a certain very vague limit, you'd have to go to the system modern languages use, forming words from a selected handful of meaningless sounds, sounds that formed a finite set but that could be combined in, for all practical purposes, an infinitude of ways.

But the point I really want to focus on is that language, like niche construction, is an autocatalytic process. Once it's started, it drives itself; it creates and fulfills its own demands. The more you do, the more you can do, and indeed the more you have to do. This may sound magical, but it really isn't. The degree of flexibility inherent in gene expression, far from limitless yet by no means negligible, interacts with the experiences of members of a species to generate new and more focused behaviors. That's how evolution works.

It's likely too that protolanguage in its later stages acquired what many might term syntax, though it really isn't. The simple fact of predication—saying something, then saying something about that thing—gives a fairly fixed serial order to utterances. In many, maybe most cases, utterances still go from the known to the unknown—they tell you a class or a name that you already know, then add some (hopefully) new information about it. So (as in the pidgin version of the barbed-weapon scenario, above) there would probably have been a statistical preponderance of what, in a true language, you'd have to call "subject-first" sentences.

But that's something you could do with beads-on-a-string chaining. To get to the next level, you had to put things together in a different way.

THE SAPLING
BECOMES AN OAK

We saw in chapter 9 that there are two ways in which words can be put together—like beads on a string or in hierarchical structures, forming A and B into [AB] and then adding C, not to A, not to B, but to the new unit [AB]. And so on, ad infinitum. Originally beads on a string was all there was. Later, much later, this way was relegated to units larger than the sentence, or to those of us who need to speak in a language barely known to us, or to initiate some means of linguistic contact from scratch (the fate of pidgin speakers worldwide). For sentences and all smaller units, the hierarchical way became universal.

When this happened, we don't know. My best guess is at the very earliest a couple of hundred thousand years ago. That's the earliest date so far suggested for the origin of our species. And it's around then that the first signs of really human behavior become manifest. Tools start to shape up a little, but it's not that. People are beginning to use ochre and other pigments to decorate their bodies (or so we assume—they were using the stuff for something, that's for sure). Types of stone used for tool manufacture are found hundreds of miles from their sources, which suggests that some form of trade had started up. That meant contact between groups that probably didn't even speak the same protolanguage.

Recall that in protolanguage, the speaker thought of a word and then transmitted it directly to the organs of speech, then the next, and the next, without linking them in the brain prior to utterance. In language today, words up to at least the phrase level are assembled within the brain and a much more complex message is sent to the vocal organs. Before that could be accomplished, at least two conditions had to be satisfied.

One we have glimpsed already, and it's essential—without it, even the simplest hierarchical structures are impossible. That's the establishment of neural links between representations of different words, representations that are widely distributed in the neocortex. (What the combined message looks like, whether it's the mere sum of two or more messages, or whether these undergo mergers and/or changes, and if so, of what kind, are mysteries that as yet we don't seem even near solving.) But there's a further condition that had to be met before hierarchical structuring could serve as a viable alternative to good old-fashioned beads-on-a-string processing.

Sending any kind of message through the brain takes up a measurable amount of time, even if it's measurable only in milliseconds. When each word goes directly to utterance on its own, that time is very short, so some serious constraints on neural messaging have little or no effect here.

But those constraints do affect longer messages. They are, one, the fact that nerves are leaky, hence the quality of any message will degrade over time, and two, the fact that the brain is a very noisy place, with all sorts of other activities going on constantly—a factor that also degrades message quality.

William Calvin of the University of Washington, the author of several popular books on human evolution, pointed out that what happens in the brain resembles what happens in the singing of choirs. If only five or six people sing together, you can tell very quickly if one of them is out of tune; if a choir of a hundred or more voices is singing, half a dozen could be out of tune and you'd never know. The variation between voices averages out, so to speak, so you hear only a single, seemingly seamless flow of sound.

In just the same way, Calvin argues, you need a large cohort of nerve cells synchronously sending out the same message if you're to override the distortion and degradation that inevitably involves some individual cells. Until you have plenty of spare cells positioned where they can be recruited to support the message, so that big choirs of neurons are all singing the same song, just building hierarchical structures can't guarantee that words or structure won't come out garbled. Until that time, it's safer, more reliable, to stick to the old beads-on-a-string routine.

Why not stick to that method anyway? Why switch to hierarchical structuring?

Language evolution, as I've said, is an autocatalytic process. It drives itself, selecting for things that will make it more effective. One of these things is sheer speed. Outside of life-or-death warning calls, speed doesn't directly affect fitness, but any organism that can get its message across sooner and get on with its life has an advantage, certainly a social advantage, over one that's slower. And haven't you ever been driven to a fury by one of those speakers who doles out his words as if he were dispensing hard-earned cash, one or two word-coins at a time? Hierarchically structured speech, as I found when I compared pidgin and creole speakers in Hawaii, is up to three times faster than beads-on-a-string speech. The first was doomed to oust the second, as soon as it became fully viable.

Linguistic versus Protolinguistic Modes

There is a story, doubtless apocryphal, about a West African state that for some reason changed from driving on the right to driving on the left (or vice versa). But drivers shouldn't worry, a government spokesman announced, because "the change will take place gradually."

Well, I'm sure the change from protolanguage took place that way (though without fatalities). However, it was a very similar kind of situation. You either drive on the left or drive on the right—there's no intermediate stage (driving down the center stripes doesn't count). In just the same way, you use either protolanguage—beads on a string—or real language—Merge with hierarchical structure. There could not have been, as some seem to suppose, a series of changes in protolanguage that brought it gradually closer to real language; either an utterance is hierarchically structured or it isn't. It was simply that more and more proto-people would use language, and those who used it would do so more and more of the time.

The situation is complicated by the fact that you can't necessarily tell whether a given utterance was produced linguistically. Take a simple sentence such as "I like chocolate." It could be structured as in (A) or (B):

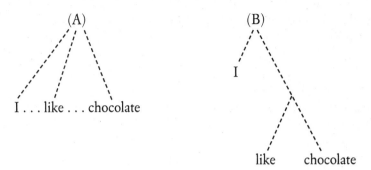

Perhaps the most important point to grasp here is that to make language the brain doesn't have to put things together in the same way as it does to make protolanguage.

If you're using protolanguage, sending each word to the speech organs the moment it bobs up, words have to be put together in the sequence in which they are uttered. There's no avoiding this. It's a logical necessity.

If you're using language, forming phrases and short clauses in the brain before uttering them, you don't have to put words together in the sequence in which they'll be uttered. In principle, you could assemble them any which way, so long as the completed phrase comes out right. In practice, it's most probable that the brain assembles sentences from the bottom up, by the simple process of first combining the words that are closest to one another.

In "I like chocolate," which are closest—"I" and "like," or "like" and "chocolate"? Well, you can put things between "I" and "like" that you can't between "like" and "chocolate"—"I sometimes like chocolate" but not "I like sometimes chocolate"—so "like" and "chocolate" are combined first, and only then is "I" combined with "like chocolate."

Note that this process—what Chomsky's minimalist program calls Merge—gives you hierarchical structure for free, so to speak. As for the linear order of spoken sentences, unavoidable if everything has to come out of one mouth, you can get this just by reading off the words at the end of each branch of the tree, from left to right.

Beads-on-a-string won't support long and complex sentences. There are several reasons why this is so. First, longer messages sent this way would take up so much time that the receiver (and maybe even the sender!) would have forgotten the beginning before the end was reached. Second, the speaker would somehow have to keep all the parts of the sentence together without the support of any brain-internal processing. Third, even assuming these two obstacles could be overcome, structural ambiguities—ambiguities due to the absence of syntax, leaving you uncertain what went with what, and where phrases and clauses began and ended—would rapidly accumulate until the processing load on the receiver became too heavy.

I grant you that in a few seconds, context and common sense between them would usually tell you what was meant. But you don't have those seconds. By the time you've used them up, the conversation's already far ahead, and you'll fall further behind with every fresh ambiguity you have to resolve. For any attempt to produce utterances longer and more complex than four- or five-word strings simply piles up the ambiguities. In language, on the other hand, structure is thoroughly predictable, and there are plenty of signals as to what that structure is. Intonation, for example: an intonation contour, a curving rise or fall or rise-fall of the voice, follows the structure of the syntax and almost always marks where boundaries between clauses lie. But you can't have a sustained intonation contour where words are popping out one at a time.

Those who clung to the protolinguistic mode would eventually have become social cripples. People already fully linguistic would have been turned off by the slowness and clumsiness of their speech, would have treated them as dummies. The big advantage hierarchical processing has over beads-on-a-string is that it's faster and also fully automatic. You don't need context, you don't need common sense; you just process the words and get the meaning instantly. The occasional ambiguities and so-called slips of the tongue (which have nothing to do with the tongue, of course, but everything to do with the degradation of neural signals described a few paragraphs ago) are a small price to pay for the immense saving in time and effort, and the ability to produce sentences of far greater subtlety and complexity than beads-on-a-string could ever achieve.

But how could hierarchical processing, in and of itself, secure such rapid and accurate comprehension?

The answer is that, in and of itself, it couldn't. It needs to be supplemented by some system of templates, something that predicts with a high degree of accuracy the kinds of thing that hierarchical structures will produce.

TEMPLATES

Simplifying somewhat for the sake of brevity and clarity, the two most crucial kinds of words are, as you might expect, nouns and verbs. And accordingly, the two templates (roughly, phrases and clauses) are headed, respectively, by nouns and verbs.

The noun template looks something like this:

$$(\text{Modifier}_n) \; \text{Noun} \; (\text{Modifier}_n) \rightarrow \text{Nmax}$$

This means that a noun may have an indefinite number of modifiers either before it, or after it, or occasionally a mix of these. Each language determines for itself on which side of the noun modifiers can go. (As in "The tall blond man with one left shoe"—English is pretty loose with its modifiers, as languages go.) The modifiers can include other phrases— [with [one left shoe]]—provided there's a preposition or something to link them with the rest of the phrase. They can even include clauses— "The tall blond man [you saw yesterday."] The total phrase we'll call Nmax, the maximal expression of a noun.

The verb template looks something like this:

$$(\text{Nmax}_n) \; \text{Verb} \; (\text{Nmax}_n) \rightarrow \text{Vmax}$$

As before, this indicates that an indeterminate number of Nmaxes can either precede or follow the verb or, again as in English, you can have it both ways. But although number must remain indefinite in a general formula, it is limited in any individual case by the number of possible arguments of the verb concerned.

What that means is that to attach to a verb, an Nmax has to fulfill a specific role relative to that verb. It has to be its Agent (whatever performs the action of the verb) or its Theme (whatever undergoes the action) or its Goal (whatever or whoever the action is directed toward). (There are other less important thematic roles, but they needn't concern us here.) Not every verb assigns the same arguments. "Fall" takes only a Theme ("Bill fell"). "Melt" can take one ("The ice [Theme] melted") or two ("Bill [Agent] melted the ice [Theme]"). A verb like "tell" can take all three ("Mary [Agent] told Bill (Goal) the time [Theme]").

Where do these templates come from? The noun template must have started to emerge the very first time two things had to be distinguished from each other: "The big one, not the small one." Soon there would be cases in which that had to be refined: "The big red one, not the small red one." The verb template is implicit in the meanings of verbs. "Fall" affects only who or what falls, so it can have only one of the three main thematic roles. "Melt" can be something that happens or something you make happen, so it may have either one or two of those roles. But if you "tell," you have to do it to somebody and you have to have something to tell them, so all three roles have to be represented somehow.

With all this apparatus, you might think you would have enough to run a language on, to compose it more or less automatically and understand it the same way, even if it was a good deal less complex than most languages are today. But wait. Where in all this is the process singled out by Chomsky as the one central and uniquely linguistic capacity—recursion?

BROUHAHA ABOUT PIRAHÃ

It's not often that a heavy-duty linguistic argument finds its way into the pages of *The New Yorker*.

However, that's exactly what happened in the spring of 2007, when that magazine featured an article on the work of Dan Everett, a linguist who for many years had been studying a language called Pirahã, spoken by an indigenous tribe in the Amazon basin.

Why would your typical *New Yorker* reader give a flying you-know-what about a language spoken by a few hundred jungle-dwelling preliterates that many even in the professional linguistic community had never heard of? In twenty-first-century terms, there's only one possible answer. The language provided data that seemed to challenge Chomsky. Challenging Chomsky is, as I mentioned in chapter 9, a continuing obsession, one that taps into the great divide between those who think culture determines the bulk of human behavior and those who think biology does it.

So the first lot thought they'd discovered a smoking gun, and for a few heady weeks the chattering classes found themselves grappling with a strange new concept: recursion.

Recursion, we are told, is the rat that ate the malt that lay in the house that Jack built. It's what enables us to expand sentences indefinitely, to infinity if need be, by inserting phrases within phrases, clauses within clauses—just like those Russian dolls that have smaller but otherwise identical dolls nesting inside them. It's what, as we saw in chapter 9, Chomsky and his colleagues regard as not only the most central part of language, but possibly the sole content of FLN, the only part that's unique to humans. Consequently, it must be a universal of human language, determined by our biological makeup—mustn't it?

But Dan Everett was claiming that Pirahā had no recursion.

Chomskyan linguists launched a massive counterattack on Everett's analysis, saying he'd gotten it all wrong, that some of his own examples disproved his claims. Arguments quickly spiraled into a technical stratosphere where few *New Yorker* readers could follow them. What hardly anyone seemed to notice was that it didn't make the slightest difference whether Everett was right or wrong.

Suppose he was right. Then the only question was, could a Pirahā baby learn a language that did have recursion? If it could—the most probable outcome—then the absence of recursion from Pirahā grammar might be rarer, but was no more remarkable, than the absence of sounds such as clicks or prenasalized consonants from English. Recursion, clicks, and prenasalized consonants are all things that human biology makes available to us. But biology doesn't mean we have to use them— once again, we're up against the myth of the gene as an unalterable, inescapable force determining our behavior down to the wire. Recursion is a more useful language component than clicks, so few if any languages manage without it, but if one human language chooses to do just that, it tells us nothing at all about the human language capacity.

But ironically, if the Pirahā baby *couldn't* learn a recursive language, that would form one of the clearest proofs of the biological nature of language anyone could ask for. It would be puzzling, because it would mean that at some stage of evolution the language capacity had branched, and consequently some things possible for folk on the major branch would be impossible for those on the minor one. But that would be the only possible explanation, because if language was cultural as some still claim, the baby (once raised in the culture of a recursive language) would surely have learned recursion anyway.

But is there really such a thing as recursion?

Even to ask the question approaches blasphemy. For half a century, everyone, whether they accepted Chomsky's theories or not, has agreed that recursion exists—that language is capable of embedding a linguistic object, a phrase or a clause, inside another linguistic object of the same kind. Whether people agreed with Chomsky or not, whether they believed that recursion was innate or not, nobody questioned that it was there, a force that had to be reckoned with.

Yet in fact, as I shall now show, it was an artifact of analysis.

Who created it? Chomsky did.

Who destroyed it? Chomsky did, only he didn't realize he had.

It's a fascinating story, and here it is.

THE STRANGE HISTORY OF RECURSION

In 1957 Chomsky published his seminal, groundbreaking work, *Syntactic Structures*. At that stage his kind of grammar was officially known as transformational-generative grammar, although, since this title was cumbersome and transformations were the most novel thing about it, most people back then called it simply transformational grammar.

Among other things, transformations took two simple sentences and made them into one complex one. Take for instance a sentence like "The girl you met yesterday speaks French." This was originally assumed to be produced by first constructing the two simple sentences, "The girl speaks French" and "You met the girl yesterday." The transformation then simply inserted the second sentence into the first, a process that came to be termed "embedding." This gave you "The girl you met the girl yesterday speaks French." The second occurrence of "the girl" was then "deleted under identity," and voilà, you had your complex sentences. All complex sentences were assumed to be constructed like this, out of simple ones.

But wait. While for heuristic or didactic purposes, transformations might be shown as operating on actual strings of words, they weren't really supposed to do this. They were actually much more abstract. Words were merely objects in "surface structure," while transformations took place at the level of "deep structure." Deep structure consisted of abstract forms, word classes and types of structure that underlay the superficial level of actual sentences. These forms were allotted symbols that were used in the instructions for transformations: S for sentence, N for noun, NP for noun phrase (since every noun was capable of expansion into a phrase), V for verb, VP for verb phrase, and so on. Sentences, down to the final transformation, were built with these terms, and words were inserted as the very last step in the sentence-forming process.

In order to construct the necessary deep structures, you needed a set of what were known as "rewrite rules." Rewrite rules broke down deep-structure labels into their constituents, as follows:

$$S \rightarrow NP\ VP$$
$$NP \rightarrow (Det)\ N\ (PP)$$
$$VP \rightarrow V\ (NP)\ (NP)\ (PP)$$
$$PP \rightarrow P\ NP$$

"Det" stands for determiner—things like "the" or "this"—and PP stands for prepositional phrase, while the presence of parentheses indicates that a constituent is optional; NP and VP need include no more than an N and a V, respectively. Adjectives were not included at this stage; sentences with adjectives, even simple sentences, were "generated" by a transformation that in order to produce "The angry man left" first had to produce "The man left" and "The man was angry," then proceed by insertion—"The man the man was angry left"—then deletion—"The man angry left"—and finally transposition.

And, as these rewrite rules showed, a unit could be included in another unit of the same kind: the NP that finished up inside a PP could subsequently be inserted inside another NP. That's recursion.

This was Chomsky's original formulation. However, it soon ran into problems that caused the theory to be revised and re-revised. First, derivations like that of "The angry man left" were abandoned, and then so were all derivations that derived complex sentences from simple ones. Complex sentences were now produced by building a "generalized phrase-marker," a string of rewrite symbols that followed (more or less) the whole outline of a complex sentence. Transformations were taken out of the sentence-building process and reserved for things like changing sentences of one type into another type—for example, active into passive, or statement into question—or moving things around in sentences, such as bringing question words to the beginning. It followed from this that clauses, allotted the same symbol as sentences (since a single clause often constitutes a sentence), had to be included in the rewrite rules, which now read:

$$S \rightarrow NP\ VP$$
$$NP \rightarrow (Det)\ N\ (PP)\ (S)$$
$$VP \rightarrow V\ (NP)\ (NP)\ (PP)\ (S)$$

So now either (or both) noun phrases and verb phrases, the constituents of sentences, could themselves contain sentences, and the picture of recursion (phrases within phrases, sentences within sentences) was complete.

However, things went on changing. The first word in the theory's title disappeared: "transformational-generative grammar" became "generative grammar" *tout court*. That was because transformations had grown fewer and fewer, until a quarter century after *Syntactic Structures* there was only one: "move alpha," which can be roughly translated as "move anything anywhere." (This might strike you as rather unhelpful, but there was by that time a series of carefully devised principles that between them sharply constrained what could move where.) But in all the time these changes were taking place, nobody stopped to look closely at any effect they might be having on the original set of assumptions that generative grammar had started out with. Terms like "recursion," "embedding," "embedded sentence," and the like now formed part of the jargon of the trade, and continued to be used by all, without anyone asking whether the newest version of the theory still licensed those terms.

By the 1990s, when Chomsky introduced his minimalist program, the original theory had changed beyond recognition. The deep structure/surface structure distinction had vanished, along with all the category labels, the NPs and VPs of the rewrite rules; those labels might sometimes still be used descriptively, as a matter of convenience, to refer to particular chunks of structure, but they no longer played any significant role in the theory. Of the elaborate network of rules and/or principles that had characterized the first versions, all that was left was the single process mentioned earlier: Merge.

Merge dealt straightforwardly with words, not category labels, and all it did was take two words and join them into a single unit. Take "Bill" and merge it with "left" and you got the sentence "Bill left." What was to stop you taking "Bill" and "right" and making the nonsentence "Bill right"? Well, the lexical properties of the words themselves told you what they needed to be attached to. "Bill" has to be attached to a verb before you can get a sentence and a verb requires a subject like "Bill"; "left" can be an intransitive verb as well as an adjective but "right" can't.

Suppose you want a longer sentence, "Bill left Mary." Do you again start by joining "Bill" and "left," then add "Mary" to give [[Bill left] Mary]? No way; sentences are built incrementally, but not that way. As we saw above, the link between verb and object, here "left" and "Mary," is much tighter than that between "Bill" and "left," so the former pair merges first.

Although this doesn't seem to have been his motive, what Chomsky has done here is provide a pretty plausible model for how brains may actually put words together to form sentences, in the real world, in real time. One reason to suppose he wasn't thinking about process is that Chomsky has always conceptualized things in terms of states rather than processes—for him, as we saw in chapter 9, almost any kind of process might just as well be instantaneous. Another is his apparent lack of interest in anything that actually happens inside brains. This is the only explanation I can think of for the all-but-incredible fact that he apparently hasn't noticed—or at least hasn't publicly admitted—what he's actually done.

By proposing Merge, he's assassinated recursion.

What really happens

Let's go back to the sentence with which we began the last section:

The girl you met yesterday speaks French.

Even though the elaborate machinery that inserted "You met the girl yesterday" inside "The girl speaks French"—a classic case of a recursive process—has long since vanished, everybody still sees this as one sentence embedded within another sentence, or to be more precise, within a noun phrase that's a constituent of the full complex sentence:

$$S^1[NP[The\ girl\ S^2[you\ met\ yesterday]S^2]NP\ speaks\ French]S^1$$

But what happens when you simply produce the sentence by successive applications of Merge?

Stage one (probably constructing in parallel, since the brain is a parallel processor): merge each verb with its complement.
    [met yesterday]   [speaks French]
Stage two: merge [met yesterday] with its subject.
    [you [met yesterday]]   [speaks French]
Stage three: merge "girl" with the product of the last merge.
    [girl [you [met yesterday]]]   [speaks French]
Stage four: close product of last merge by merging determiner.
    [the [girl [you [met yesterday]]]   [speaks French]
Stage five: Merge the two constituents.
    [[the [girl [you [met yesterday]]]] [speaks French]]

Where was anything inserted or embedded in anything else? Nowhere. Words were simply merged with other words in a purely additive process. It is only when we look at the finished product and start to add category labels to it that *in retrospect it looks as if* one constituent has been placed inside another. But in the actual process of sentence formation, nothing you could call recursion has taken place.*

Contra what Chomsky has claimed and most people have assumed, there's no special ability to deal with recursive processing that has somehow evolved in the human species. There is, to the contrary, a complete absence of rules and restrictions. You can merge anything to form a sen-

---

*Under pressure from the above argument, generativists such as Luigi Rizzi have moved to a weaker definition of recursion: "Any process that takes the output of one step as the input to the next." If we accept this definition, Merge itself becomes a recursive process. But then so are many behaviors regularly produced by nonhuman species—for example, birds building nests (step one, interweave two twigs; step two, weave a third twig into the original two; step three, weave a fourth twig into the preceding three . . .). Generativists thus find themselves in the following dilemma: they must either accept the first definition and admit that recursion plays no part in syntax, or accept the second and admit that no part of syntax is unique to humans.

tence—words, phrases, clauses, you name it—provided that all the lexical requirements of all the words you merge are satisfied and none of them remain unsatisfied. Because what you are merging are actual words, or mergers of words, not abstract categories or chunks of labeled structure. It's precisely the absence of any restriction on what type of object can be merged that allows the illusion of recursive processing to exist.

Please note, I'm not taking any credit for any kind of great new discovery here. All I've done is follow through on the logic of what was actually done by Chomsky himself—*il miglior fabbro,* "the better craftsman," as T. S. Eliot famously said of Ezra Pound. The process of Merge, when applied, rules out any necessity for supposing that language requires recursion. Yet, as we saw in chapter 9, it was Chomsky, along with coauthors Hauser and Fitch, who tried to send language evolutionists wild-goose-chasing off through all the highways and byways of evolutionary biology in search of the Holy Grail, the mysterious capacity perhaps somehow used by some other species for numbering, or navigation, or social interaction, or . . .

THE FULL FLOWERING

Of course, what I've covered here is very far from a full account of language, even of grammar. We've reached the stage where it became possible to build merged, hierarchical sentence structures. A lot more would follow. There's inflection, and agreement, and case marking, and empty categories, and anaphoric relationships, and much, much more. But with the ingredients I've described, you can get at least the skeleton of a full human language up and running.

And once that had happened, there seemed no limit to what could be accomplished. With the full force of a symbolic-syntactic language at its service, our species began to turn out novel artifacts. Slowly at first, since there are many steps that can't be taken unless they're preceded by certain other steps.

That's a part of what explains the apparent time lag between the emergence of the human species and the flowering of creativity that occurred when that species arrived in Europe. But only a part; there's also the fact that what looks like the Great Leap Forward is largely a sampling error. The vast majority of archaeological sites so far investigated lie inside Europe; as more and more African sites are opened up, the picture changes, and human innovations appear at earlier and earlier dates. And there's a third, perhaps even more potent factor involved.

Humans tend to embody contradictory traits. We are highly cooperative yet highly competitive. We are strongly innovative yet staunchly conservative. It's the second contradiction that applies here. There's a strong tendency to hold on to what we know, to fear change rather than desire it. And remember, biological developments don't mandate new behaviors—they merely make them possible. Whether those possibilities are exercised is a matter of choice, entirely up to us.

"War is the locomotive of history," Trotsky said. Just as World War II called forth all the energy and ingenuity of the English and Americans, so did the conflict with a species of almost equivalent abilities—the Neanderthals—call forth all the energy and ingenuity of the Cro-Magnons. That, rather than any mutation, any sudden surge in capacity, is most likely what accounts for the Great Leap Forward.

After that, the construction of new niches developed at an unprecedented rate. First the herding niche, then the agricultural niche, finally the industrial niche, as humans struggled to adapt the world to themselves—first controlling other animals, then plants, then energy and matter themselves. Was there no limit?

Of course there was. Of the types of niche construction that John Odling-Smee and his colleagues described, one was negative niche construction. By exhausting the capacity of a niche, or by choking on the debris caused by construction, a species could construct itself into extinction. How close to that we've already come is anyone's guess.

But in the process humans set in motion, a stranger fate lurked.

FROM APE TO ANT?

Imagine the following scenario.

You're looking at a split screen, with different videos showing simultaneously on each side. The video on the left shows an anthill with its top removed so as to reveal the network of tunnels beneath, and the ants moving to and fro along those tunnels. The video on the right, taken from a much greater height, shows a human city with its network of streets, and the humans moving to and fro along those streets. What you see in each is a multitude of small dark objects moving constantly and rapidly around, seemingly at random, with no apparent purpose but yet with a vigor and intensity that reeks of purposeful action. What are they all doing, you wonder?

I can think of no comparable display, featuring two species as phylogenetically remote from each other as we are from ants, where the images on the two screens would so closely resemble each other. Is this just a freakish accident, or does it contain within it some deep and disturbing truth?

Ants have already cropped up in our story. Adoption of an antlike form of subsistence—the scavenging of carcasses far larger than ourselves—seems likelier than any other proximate cause to have given rise to the birth of language. But as we developed, more and more aspects of our existence came to resemble those of ants.

Our numbers increased to antlike size. From the few hundred thousand, or maybe million or two—typical mammalian-species numbers—who inhabited the earth less than a thousand lifetimes ago, our population ballooned, with ever-increasing speed, to numbers that before had been achieved only by insects. Just as ants domesticated aphids, pasturing them on plants, stroking them until they exuded fluids, so did we domesticate cattle, pasturing them on grass, milking them. Just as ants prepared beds, planted spores, brought in plant food, and harvested the resulting fungi, so did we prepare fields; plant seeds; fertilize, compost, and manure them; and harvest the resulting cereals and other crops. Just as ants built enormous underground cities, so did we build enormous aboveground cities. Are all these things merely coincidences?

Of course not. Niche construction processes determine the kind of occupations a species will follow and the kind of society it will have to live in as a result. Whether the niche is created slowly, by instinct, over millions of years or (in part at least) by cultural learning over mere thousands makes no difference. The niche makes the difference. The only question is, are we through yet, or is it still changing us?

Nonsense, the humanist answers. We're free, independent souls, above the laws that rule the rest of creation. Poppycock, says the orthodox biologist. We're just a species of primate, full of those good old primate genes—a gentrified ape, true, but one still too ornery to succumb to an ant fate.

Wait a minute. There was a time when ants too were free-roving organisms. It happened to them; why can't it happen to us? The degree of social control under which we already labor would have been both incomprehensible and intolerable to our hunting-and-gathering ancestors. Why is it, do you suppose, that when a hunter-gatherer group is sucked into the vortex of "civilization," so many of its members seem to undergo a kind of spiritual death, quickly falling victim to drugs, alcohol, irrational violence, or suicidal despair? Think about it.

And think about this: for ten thousand years, ever since cities and government began, we have been selecting against the most independent, individualistic members of our species. Rebels, revolutionaries, heretics, criminals, martyrs—all those opposed to the current norms of society—have been systematically imprisoned, exiled, murdered, or executed throughout the last hundred centuries. Since the vast majority died young or spent their procreative years in monosexual jails, their contribution to the human gene pool has been negligible. But the passive, the compliant, the loyal, the obedient—they prospered like the green bay tree, spreading their seed far and wide. Has this really had no effect on human nature?

I used to think, in common with most people, that (apart from oddities like lactose tolerance or sickle-cell anemia) evolution in the human species was effectively over. In the last few years we've learned that this is not the case. Evolution is proceeding, genes are changing, in ways we still can't fully understand. By the time we understand them, the damage may have been done. It doesn't take many generations to turn a wolf into a dog.

Already there have been signs and portents. During the past couple of thousand years, caste systems—systems like those of ants, where an individual's occupation and fate are predestined at birth—have come into existence in many parts of the world, most strikingly in India. To most of us, caste systems are just quaint and rather repellent aberrations, quirks of history swamped now in a rising world tide of democracy. I'm inclined to suspect that this view may be dangerously optimistic. They may instead be better seen as trial runs, premature precursors of what is to come once the last few kicks in our ape nature have been eliminated.

At least, that's something worth starting to think about.

There is one consolation. The path of runaway niche construction moves with a powerful current, but not necessarily an undivertable one. The very notion of niche construction asserts the autonomy of the organism, the power latent in species to influence their own destiny. Our niche gave us language, language gave us intelligence, but only the wise use of that intelligence can keep us free and fully human.

NOTES

INTRODUCTION

3 *gestures hearing people use*: Frishberg 1987, Torigoe and Takei 2002.
5 *Look up "human being"*: www.britannica.com/EBchecked/topic/275376/human-being.
5 *"If it be maintained"*: Darwin 1871, p. 330.
6 *the psychologist Eric Lenneberg*: Lenneberg 1967.
6 *"the hardest problem in science"*: Christiansen and Kirby 2003.
8 *I have written elsewhere*: Bickerton 2008.
8 *"human animals"*: Penn et al. 2008.
9 *flight in insects*: Pringle 1975, Sane 2003.
10 *"Adaptation is always asymmetrical"*: Williams 1992.
10 *"selfish gene"*: Dawkins 1976.
10 *just too many things*: Johansson 2005, especially chapter 11.
10 *the "primate-centric" approach*: Pepperberg 2005.
11 *But that environment*: Odling-Smee et al. 1996, 2003.
13 *people who talk about "precursors"*: e.g., Pollick and de Waal 2007, Hurford 2007.

1: THE SIZE OF THE PROBLEM

16 *Marc Hauser published*: Hauser 1996.
17 *the earliest ethologists*: Lorenz 1937, Tinbergen 1963, Krebs 1991.
18 *the sounds chimps make*: According to Michael Wilson (www.discoverchimpanzees.org/activities/sounds_top.php), "The precise number of call types is difficult to determine, since many calls grade into one another, producing intermediate forms"—something that is never the case with words.
19 *Now researchers have found*: Cheney and Seyfarth 1990.
19 *As Darwin long ago*: Darwin 1871.
20 *the whites of our eyes*: Tomasello 2007.
20 *"imagine what might happen"*: Pinker 1994, p. 333.
24 *ranges overlap*: Wilson 1972.
24–25 *chimpanzees*: Gardner and Gardner 1969, Terrace 1979; *gorillas*: Patterson and Linden 1981; *bonobos*: Savage-Rumbaugh et al. 1986, Savage-Rumbaugh and Lewin 1994;

orangutans: Miles 1990; *bottlenose dolphins*: Herman et al. 1984, Herman and Forestell 1985; *African gray parrots*: Pepperberg 2000; *sea lions*: Schusterman and Krieger 1984.

26 *chimps on the Ivory Coast*: Boesch and Boesch 1990.
26 *the last common ancestor was*: Chen and Li 2001.
26 *when chimpanzees were observed*: Boesch 1994.
26 *social intelligence*: Humphrey 1976, Povinelli 1996, Worden 1998.
26 *"Machiavellian strategies"*: Byrne and Whiten 1988.
27 *"grooming and gossip" theory*: Dunbar 1996.
29 *female choice*: Miller 1997.
29 *if you trim a peacock's tail*: Alcock 2001, p. 348.
29 *John Wilkes*: Sainsbury 2006.
31 *the harder a signal is to fake*: Zahavi 1975, 1977. (For a critical view see Maynard Smith 1976.)
33 *"in order to survive"*: Jablonski 2007.
33 *three levels of intelligence*: Macphail 1987.
35 *the continuity paradox*: Bickerton 1990, p. 8.
36 *an engineering perspective*: Hauser 1996, pp. 638–52.

2: THINKING LIKE ENGINEERS

38 *pidgins and creoles*: Bickerton 2008, Arends et al. 1994.
39 *not necessarily a good model*: Slobin 2001.
40 *protolanguages*: Westcott 1976.
40 *a Broca's aphasic*: Goodglass and Geschwind 1976.
42 *Diana monkeys*: Zuberbühler 2002, 2005.
43 *The vervets' calls*: Cheney and Seyfarth 1990.
45 *"The Human Revolution"*: Hockett and Ascher 1964, 1992.
47 *An indexical sign*: Peirce 1978, Deacon 1997, chapter 3.
49 *marriage?!*: Deacon 1997, pp. 402–407.
50 *"probably not until Homo erectus"*: ibid., p. 407.
50 *"displacement"*: Pearce 1997, p. 258.
52 *An iconic sign*: Armstrong 1983.
52 *According to him*: Deacon 1997, pp. 69–79.

3: SINGING APES?

55 *popular works on human evolution*: Morris 1999, Diamond 1992, McCrone 1992, Burling 2005.
56 *"And yet it is"*: Mithen 2005, p. 113.
57 *In the Shadow of Man*: Goodall and von Lawick 2000.
59 *ACSs of bonobos and chimpanzees*: Estes 1991, Pollick and de Waal 2007.
60 *"food-peep" vocalization*: Krunkelsven et al. 1996.
62 *it appeared "probable"*: Darwin 1871, p. 573.
62 *"language was born"*: Jesperson 1922, p. 434.
62 *"musilanguage"*: Mithen 2005.
62 *a form of sexual display*: Miller 1997, 2000.
62 *Gibbons*: Deputte 1982.

62 *aquatic ape hypothesis*: Morgan 1982.

63 *several main functions*: Geisemann 2000.

65 *the idea of protolanguage*: Bickerton 1990, chapter 5.

66 *another highly appealing idea*: Wray 1998, 2000, Arbib 2008.

68 *there'd be the problem*: Tallerman 2007, 2008.

70 *Which presents a problem*: Falk 2004.

71 *In my commentary*: Bickerton 2004.

4: CHATTING APES?

73 *"strange creature . . . from Guiny"*: Pepys 2000, p. 160.

73 *their chimpanzee Viki*: Hayes and Nissen 1971.

74 *Wild Boy of Aveyron*: Candland 1993.

74 *"the one great barrier"*: Müller 1870.

75 *Clever Hans*: Pfungst 1911.

76 *"Washoe learned"*: Gardner and Gardner 1978, p. 73; *"Apes appear"*: Miles 1978, p. 114;
   "Koko has learned": Patterson 1985, p. 1.

78 *Kanzi scored correctly*: Savage-Rumbaugh et al. 1993.

81 *Lana, an ape*: Rumbaugh 1977.

82 *"Neurons that fire together"*: Hebb 1949.

84 *a nice distinction*: Számadó and Szathmáry 2006.

85 *Sea lions*: Schusterman and Krieger 1984; *dolphins*: Herman 1986; *parrot*: Pepperberg 2000.

87 *never say "higher" or "lower"*: Darwin, cited in Mayr 1982, p. 367.

88 *their brains are configured*: Jarvis and Mello 2000, Striedter 1994.

89 *ape research facility*: www.iowagreatapes.org/index.php.

89 *bonobos*: de Waal 1988, 1995, 1997, Kano 1992.

5: NICHES AREN'T EVERYTHING (THEY'RE THE ONLY THING)

92 *"Nothing in biology"*: Dobzhansky 1964, p. 449.

92 *"Adaptation is always"*: Williams 1992, p. 484.

93 *Beavers*: Muller-Schwarze and Sun 2003.

95 *Lamarck*: Packard 1901.

97 *98 percent of Swedes*: Simoon 1969.

98 *Darwin's Dangerous Idea*: Dennett 1996.

98 *niche construction theory*: Odling-Smee et al. 2003.

99 *importance of behavior in evolution*: Waddington 1969, Lewontin 1983, Dawkins 1982.

99 *"an animal's behaviour"*: Dawkins 1982, p. 233.

100 *"the ecological niche"*: Odum 1959.

100 *earthworms*: Lee 1985, Satchell 1983, Darwin 1881.

103 *"pernicious"*: Dawkins 2004.

104 *Japanese macaque monkeys*: Kawai 1965; *Ivory Coast*: Boesch and Boesch 1990.

105 *"While other species"*: Bickerton 1990, p. 232.

105 *termites*: Lüscher 1961, Abe et al. 2000; *leaf-cutter ants*: Wilson 1980.

6: OUR ANCESTORS IN THEIR NICHES

110 *"Over time, genetic change can alter"*: encarta.msn.com/text_761566394_1/human_evolution.html.

110 *FOXP2*: Marcus and Fisher 2003.

110 *pleiotropic genes*: Caspari 1952, Williams 1957.

111 *a very simple niche distinction*: Wrangham and Peterson 1998; Estes 1991.

112 *the climate began to change*: deMenocal 1995.

112 *australopithecines*: Dart 1925.

112 *They had big teeth*: Wolpoff 1973, Walker 1981.

113 *Man the Hunted*: Hart and Sussman 2005.

114 *names are enough to induce fear*: Carroll 1988, Turner 1997.

114 *the Taung child*: Berger and Clark 1995.

116 *"functional reference"*: Dittus 1984, Cheney and Seyfarth 1988, Hauser 1998.

117 *the overall drying trend*: Reed 1997.

118 *ambush hunting*: Chazan and Horwitz 2006.

118 *endurance hunting*: Bramble and Lieberman 2004.

120 *Australopithecus garhi*: Asfaw et al. 2000.

120 *bone marrow*: Cordain et al. 2001.

120 *cutmarks made by primitive tools*: Semaw et al. 2003.

121 *brains began to grow*: Lee and Wolpoff 2003.

122 *the size niche*: Case 1979.

122 *Nicholas Toth*: Schick and Toth 1993.

123 *"Initially, the sight"*: ibid., p. 166.

123 *"catchment scavenging"*: Binford 1985, Ulijaszek 2002.

123 *"territory scavenging"*: Binford 1985, Blumenschine 1991, Larick and Ciochon 1996.

124 *the African elephant*: Species Survival Commission, African Elephant Specialist Group's 2007 status report: data.iucn.org/themes/ssc/sgs/afeg/aed/aesr2007.html.

125 *sequences of cut marks*: Blumenschine 1987, Blumenschine et al. 1994, Monahan 1996, Domínguez et al. 2005.

126 *optimal foraging theory*: Stephens and Krebs 1986, Schmitz 1992, Irons et al. 1986, Velasco and Millan 1998.

127 *Nathan Bedford Forrest*: Catton 1971.

7: GO TO THE ANT, THOU SLUGGARD

128 *an article coauthored*: Hauser, Chomsky, and Fitch 2002.

128 *"Current thinking in neuroscience"*: ibid, p. 1572.

129 *"evo-devo"*: Goodman and Coughlin 2000, Carroll 2005.

129 *"Much that has been learned"*: Mayr 1963, p. 609.

130 *the mouse and the fly*: Müller et al. 1995, Thor 1995.

131 *ACS of bees*: Frisch 1967; Gould 1976; Dyer and Gould 1983.

132 *eusocial*: Wilson 1971, Gadagkar 1990.

135 *ants are extractive foragers*: Wilson 1962, Hangartner 1969, Moglich and Hölldobler 1975, Hölldobler 1978.

135 *release a chemical*: Allies et al. 1986.

137 *he removed the lead ant*: Hölldobler 1971.

138 *"When a forager finds"*: Sudd and Franks 1987, p. 112.

139 *Ravens in Winter*: Heinrich 1991.
142 *did culture seemingly stagnate*: Hadingham 1980.
143 *Acheulean hand ax*: O'Brien 1981, Calvin 1993, Davidson and Noble 1993, Kohn and Mithen 1999.

8: THE BIG BANG

147 *"Out of Africa"*: Stringer and McKie 1996.
147 *multiregional hypothesis*: Thorne and Wolpoff 1992.
147 *As the human family multiplied*: compare, for example, anthropology.si.edu/humanorigins/ha/a_tree.html and www.livescience.com/history/070831_hn_family_tree.html.
149 *Microevolution's fine by them*: www.antievolution.org/features/nas_ohio_20040209.pdf.
149 *speciation*: Foley and Lahr 2005.
149 *the published proceedings*: Mellars et al. 2007.
149 *human and chimp ancestors*: Disotell 2006.
150 *Ilya Ivanov*: journals.cambridge.org/action/displayAbstract?fromPage=online&aid=124129.
151 *a map showing archaeological sites*: www.handprint.com/LS/ANC/disp.html.
151 *recent discoveries*: Spoor et al. 2007.
151 *"The fact that they stayed separate"*: Maeve Leakey, quoted in *The Washington Post*, Science Notebook, August 13, 2007.
153 *"variable speedism"*: Dawkins 1987, p. 247.
154 *continental drift*: Wegener 1924.
156 *woman the gatherer*: Dahlberg 1975.
157 *man the hunter*: Lee and DeVore 1968, Stanford 1999.
157 *"Did bands of early humans"*: Stanford 1999, p. 106.
159 *ants are eusocial*: Wilson 1971.
160 *Saramaccan*: Price 1976.
161 *men hunt and women gather*: Panter-Brick et al. 2001, Barnard 2004.
162 *the enormous number of hand axes*: Isaac 1977, Schick 2001.
162 *projectiles used in hunting*: O'Brien 1981, Calvin 1993.
162 *a form of sexual display*: Kohn and Mithen 1999, Miller 2000.
163 *hacking out chunks*: Bunn and Kroll 1986.
164 *"Neither would transport"*: O'Connell et al. 1999, p. 478. See also O'Connell et al. 1988.
167 *"Our Miocene primate ancestors"*: Boyd and Richerson n.d., p. 4.
168 *"small variations of the"*: Wikipedia, "Butterfly effect" (en.wikipedia.org/wiki/Butterfly_effect).

9: THE CHALLENGE FROM CHOMSKY

169 *he trounced B. F. Skinner*: Chomsky 1959.
170 *twin demons of innatism*: Hauser 1996, 33–43.
170 *"Well, it seems to me"*: Chomsky cited in Harnad et al. 1976.
170 *ELIZA program*: Weizenbaum 1966.
171 *the 2002 paper*: Hauser, Chomsky, and Fitch 2002.
171 *"language, as good a trait"*: Hauser 1996, p. 32.
171 *"it is almost universally"*: Chomsky 1968, p. 59.

172 *"fits beautifully with the conceptual"*: Hauser 1996, p. 49.
172 *"to attribute this development"*: Chomsky 1972, p. 97.
174 *thirteen properties*: Hockett 1960.
175 *cotton-top tamarins*: Ramus et al. 2000.
176 *inability to distinguish those sounds*: Kojima 1990.
176 *biological program for language*: Bickerton 1981, 1984.
176 *twice as many genes as fruit flies*: www.sciencedaily.com/releases/2007/12/071214094106.htm.
177 *The journal* Nature: Bickerton 1996.
177 *"an overarching concern"*: Hauser, Chomsky, and Fitch 2002, p. 1572.
179 *"recursion in animals"*: ibid, p. 1578.
179–80 *Steven Pinker and Ray Jackendoff*: Pinker and Jackendoff 2005, Jackendoff and Pinker 2005.
180 *nine stages*: Jackendoff 2002.
180 *in a 1990 paper*: Pinker and Bloom 1990.
181 *In some small group*: Chomsky 2005.
181 *There do, however*: Chomsky 2005. See also Huybregts 2006.
184 *thought and communication*: Bickerton 1990.
185 *gossip as the engine*: Dunbar 1996.
185 *Fibonacci numbers*: Jenkins 2000.
185 *idealization of instantaneity*: Botha 1999.
186 *"unbounded Merge"*: Chomsky 1995.
187 *Principle of associativity*: Perry et al. 1988, p. 364.

10: MAKING UP OUR MINDS
192 *"takes information from"*: Marcus 2004, p. 114.
194 *"offline thinking"*: Bickerton 1995.
195 *"Ever since Darwin"*: Penn et al. 2008. (For an orthodox view, see Premack 2004.)
196 *"but otherwise cognitively"*: Hurford 2007, p. 164.
196 *"for over 35 years"*: Pepperberg 2005.
197 *New Caledonian crow*: Hunt and Gray 2003.
198 *experiments with pigeons*: Herrnstein 1979.
199 *Scrub jays*: Clayton and Dickinson 1998, Clayton et al. 2001.
200 *a series of experiments*: Zuberbühler et al. 1999.
200 *"On hearing first"*: Hurford 2007, p. 227.
201 *those categories didn't jell*: Langer 2006.
202 *David Premack showed*: Premack 1983.
203 *Japanese macaques*: Kawai 1965, Kawamura 1959.
203 *Bears rifle garbage cans*: Pitt and Jordan 1996.
203 *Aterian points*: Shea 2006.
204 *difference between a concept and a category*: Lambert and Shanks 1997, Langer 2006.
206 *different kinds of memory*: for episodic memory, Tulving 2002, Suddendorf and Busby 2003; for semantic memory, Martin and Chao 2001; for procedural memory, Tamminga 2000.

11: AN ACORN GROWS TO A SAPLING

211 *twice as big as those of apes*: McHenry 1994, table 1.
212 *Long-continued increase*: Tobias 1971.
212 *The fossil and archaeological record*: Falk 1993, p. 226.
213 *more than doubled, in Neanderthals*: Evans et al. 2005.
213 *For more than a million years*: Jellinek 1977.
214 *hardly more complex than those of Cro-Magnons*: Fagan and Van Noten 1971, Best 2003.
217 *arbitrariness*: Saussure 2006.
220 *between 2 million and 1.6 million years*: Roche et al. 2002, Monahan 1996, Larick and Ciochon 1996.
221 *new and significant information*: Dessalles 2008.
222 *they don't negate like mother does*: Brown 1973.
223 *figment*: Dennett 1991, p. 346.
225 *the good old-time pidgins*: Bickerton 1981, 2008.
225 *"almost unimaginable monotony"*: Jellinek 1977, p. 28.
228 *"In the forest, Fred"*: Bowie 2008.
230 *Rotokas:* Robinson 2006; *Ixoo:* Traill 1981.

12: THE SAPLING BECOMES AN OAK

232 *beginning to use ochre*: Marshack 1981.
232 *some form of trade*: McBrearty and Brooks 2000, Feblot-Augustine 1998.
233 *establishment of neural links*: Pulvermuller 2002.
233 *degrades message quality*: Calvin and Bickerton 2000.
236 *follows the structure of the syntax*: Dogil et al. 2002.
237 *Each language determines*: Croft and Deligianni 2001.
237 *fulfill a specific role*: Pinker 1989, Grimshaw 1990.
238 *Pirahã*: Everett 2005, 2007.
238 *New Yorker*: Colapinto 2007.
239 *Chomskyan linguists launched*: Nevins et al. 2007.
240 *Syntactic Structures*: Chomsky 1957.
242 *"generalized phrase-marker"*: Chomsky 1965.
244 *Generalivists such as Luigi Rizzi*: Rizzi 2009.
245 il miglior fabbro: Eliot 1998, p. 53.
246 *the Great Leap Forward*: Klein 2002, McBrearty and Brooks 2000, d'Errico et al. 2005.
246 *negative niche construction*: Odling-Smee et al. 2003, Diamond 2005.
248 *this is not the case*: Balter 2005, Voight et al. 2006.

BIBLIOGRAPHY

Abe, Takuya, David Edward Bignell, and Masahiko Higashi, eds. 2000. *Termites: Evolution, sociality, symbioses, ecology.* New York: Springer.

Aboitiz, Francisco, and Ricardo Garcia. 1997. The evolutionary origin of the language areas in the human brain: A neuroanatomical perspective. *Brain Research Reviews* 25:381–96.

Alcock, J. 2001. *Animal behavior.* 7th ed. Sunderland, Mass. Sinauer Associates.

Allies, A. B., A.F.G. Bourke, and N. R. Franks. 1986. Propaganda substances in the cuckoo ant *Leptothorax kutteri* and the slave-maker *Harpagoxenus sublaevis. Journal of Chemical Ecology* 12:1285–93.

Arbib, Michael A. 2008. Holophrasis and the protolanguage spectrum. *Interaction Studies* 9:151–65.

Arends, Jacques, Pieter Muysken, and Norval Smith, eds. 1994. *Pidgins and creoles: An introduction.* Amsterdam: Benjamins.

Armstrong, David F. 1983. Iconicity, arbitrariness, and duality of patterning in signed and spoken language. *Sign Language Studies* 38:51–69.

Asfaw B., T. White, O. Lovejoy, B. Latimer, S. Simpson, and G. Suwa. 2000. Australopithecus garhi: a new species of early hominid from Ethiopia. *Science* 284: 629–35.

Balter, M. 2005. Are humans still evolving? *Science* 309:234–37.

Barnard, A. J., ed. 2004. *Hunter-gatherers in history, archaeology and anthropology.* Oxford, U.K.: Berg.

Berger, L., and R. Clark. 1995. Eagle involvement in accumulation of the Taung child fauna. *Journal of Human Evolution* 29:275–99.

Bermúdez, J. 2003. *Thinking without words.* Cambridge, Mass.: MIT Press.

Best, Ann. 2003. *Regional variation in the material culture of hunter-gatherers.* Oxford, U.K.: British Archaeological Reports International Series 1149.

Bickerton, Derek. 1981. *Roots of Language.* Ann Arbor, Mich.: Karoma.

———. 1984. The language bioprogram hypothesis. *Behavioral and Brain Sciences* 7: 173–221.

———. 1990. *Language and species.* Chicago: University of Chicago Press.

———. 1995. *Language and human behavior*. Seattle: University of Washington Press.

———. 1996. Chattering classes: Review of Hauser 1996. *Nature* 382:592–93.

———. 2005. Language evolution: A brief guide for linguists. *Lingua* 117:510–26.

———. 2008. Darwin's last word (and "word" is le mot juste): Commentary on Penn et al. 2008.

Binford, L. S. 1985. Human ancestors: Changing views of their behavior. *Journal of Anthropological Archaeology* 4:292–327.

Blumenschine, R. J. 1987. Characteristics of an early hominid scavenging niche. *Current Anthropology* 28:383–407.

———. 1991. Hominid carnivory and foraging strategies, and the socio-economic function of early archaeological sites. *Philosophical Transactions of the Royal Society of London* 334B:211–21.

Blumenschine, R. J., J. A. Cavallo, and S. P. Capaldo. 1994. Competition for carcases and early hominid behavioral ecology. *Journal of Human Evolution* 27:197–214.

Boesch, C. 1994. Hunting strategies of Gombe and Tai chimpanzees. In *Chimpanzee cultures*, ed. R. W. Wrangham, Cambridge, Mass.: Harvard University Press, pp. 77–92.

Boesch, C., and H. Boesch. 1990. Tool use and tool making in wild chimpanzees. *Folia Primatologia* 54:86–99.

Botha, R. 1999. On Chomsky's "fable" of instantaneous language evolution. *Language and Communication* 19:243–57.

Bowie, Jill. 2008. Proto-discourse and the emergence of compositionality. *Interaction Studies* 9:18–33.

Boyd, R., and P. J. Richerson. n.d. Solving the puzzle of cooperation. Ms., University of California, Los Angeles. Available at www.sscnet.ucla.edu/anthro/faculty/boyd/Fyssen99.pdf.

Bramble, D. M., and D. E. Lieberman. 2004. Endurance running and the evolution of *Homo*. *Nature* 432:345–53.

Brodie, D. E., III. 2005. Caution: niche construction ahead. *Evolution* 59:249–51.

Brown, Roger. 1973. *A first language*. Cambridge, Mass.: Harvard University Press.

Bunn, H. T. and E. M. Kroll. 1986. Systematic butchery by Plio-Pleistocene hominids at Olduvai Gorge, Tanzania. *Current Anthropology* 27:431–52.

Burling, Robbins. 2005. *The talking ape: How language evolved*. Oxford: Oxford University Press.

Byrne, Richard W., and Andrew Whiten. 1988. *Machiavellian intelligence, social expertise, and the evolution of intellect in monkeys, apes and humans*. Oxford, U.K.: Clarendon Press.

Calvin, W. 1993. The unitary hypothesis: A common neural circuitry for novel manipulations, language, plan-ahead and throwing. In *Tools, language and cognition in human evolution*, ed. K. R. Gibson and T. Ingold, 230–50. Cambridge: Cambridge University Press.

Calvin, William, and Derek Bickerton. 2000. *Lingua ex machina*. Cambridge, Mass.: MIT Press.

Candland, Douglas K. 1993. *Feral children and clever animals*. Oxford: Oxford University Press.

Caramazza, Alfonso. 1996. The brain's dictionary. *Nature* 380:485–505.

Carroll, R. L. 1988. *Vertebrate paleontology and evolution*. New York: Freeman.

Carroll, S. B. 2005. *Endless forms most beautiful: The new science of Evo-Devo and the making of the animal kingdom*. New York: Norton.

Case, T. J. 1979. Optimal body size and an animal's diet. *Acta Biotheoretica* 28:54–69.

Caspari, E. 1952. Pleiotropic gene action. *Evolution* 6:1–18.

Catton, Bruce. 1971. *The Civil War*. New York: American Heritage Press.

Chater, N., and C. Heyes. 1994. Animal concepts: Content and discontent. *Mind and Language* 9:209–46.

Chazan, Michael, and Liora Kolska Horwitz. 2006. Finding the message in intricacy: The association of lithics and fauna on Lower Paleolithic multiple carcass sites. *Journal of Anthropological Archaeology* 25:436–47.

Chen, F., and W. Li. 2001. Genomic divergences between humans and other hominoids and the effective population size of the common ancestor of humans and chimpanzees. *The American Journal of Human Genetics* 68(2):444–56.

Cheney, Dorothy, and Robert Seyfarth. 1988. Assessment of meaning and the detection of unreliable signals by vervet monkeys. *Animal Behavior* 36:477–86.

———. 1990. *How monkeys see the world*. Chicago: University of Chicago Press.

Chomsky, Noam. 1957. *Syntactic structures*. The Hague: Mouton.

———. 1959. A review of B. F. Skinner's *Verbal behavior*. *Language* 35:26–58.

———. 1965. *Aspects of the theory of syntax*. Cambridge, Mass.: MIT Press.

———. 1972. *Language and mind*. Enlarged edition. New York: Harcourt Brace Jovanovich.

———. 1995. *The minimalist program*. Cambridge, Mass.: MIT Press.

———. 2005. Some simple evo-devo theses: How true might they be for language? Paper presented at the Morris Symposium on the Evolution of Language, Stony Brook, N.Y., October 2005.

Christiansen, M. H., and S. Kirby. 2003. Language evolution: The hardest problem in science? In *Language evolution*, ed. M. H. Christiansen and S. Kirby, 1–15. Oxford: Oxford University Press.

Clayton, N. S., and A. Dickinson. 1998. Episodic-like memory during cache recovery by scrub jays. *Nature* 395:272–74.

Clayton, N. S., D. Griffiths, N. Emery, and A. Dickinson. 2001. Elements of episodic-like memory in animals. *Philosophic Transactions of the Royal Society of London* B356:1483–91.

Colapinto, John. 2007. The interpreter: Has a remote Amazonian tribe upended our understanding of language? *The New Yorker*, April 16, 2007.

Cordain, L., B. A. Watkins, and N. J. Mann. 2001. Fatty acid composition and energy density of foods available to African hominids. *World Review of Nutrition and Dietetics* 90:144–61.

Croft, William, and Efrosini Deligianni. 2001. Asymmetries in NP word Order. Paper presented at the International Symposium on Deictic Systems and Quantification in Languages Spoken in Europe and Northern and Central Asia, Udmurt State University, Izhevsk, Russia, May 2001.

Dahlberg, Frances. 1975. *Woman the Gatherer*. New Haven, Conn.: Yale University Press.

Damasio, H., T. J. Grabowski, D. Tranel, R. D. Hichwa, and A. R. Damasio. 1996. A neural basis for lexical retrieval. *Nature* 380:499–505.

Damasio, A. R., and H. Damasio. 1992. Brain and language. *Scientific American* 267(3): 89–95.

Dart, Raymond. 1925. Australopithecus africanus: The man-ape of South Africa. *Nature* 115:195–99.

Darwin, Charles. 1871. *The descent of man, and selection in relation to sex.* New York: Appleton and Co.

———. 1881. *The formation of vegetable mold through the action of worms, with observations on their habits.* London: John Murray.

Davidson, I., and W. Noble. 1993. Tools and language in human evolution. In *Tools, language and cognition in human evolution,* ed. K. R. Gibson and T. Ingold, 230–50. Cambridge: Cambridge University Press.

Dawkins, R. 1976. *The selfish gene.* Oxford, U.K.: Oxford University Press.

———. 1982. *The extended phenotype.* Oxford, U.K.: Freeman.

———. 1987. *The blind watchmaker.* New York: Norton.

———. 2004. Extended phenotype—but not too extended: A reply to Laland, Turner and Jablonka. *Biology and Philosophy* 19:377–96.

Deacon, Terrence. 1997. *The Symbolic Species.* New York: Norton.

deMenocal, Peter B. 1995. Plio-Pleistocene African climate. *Science* 270:53–59.

Dennett, Daniel C. 1991. *Consciousness explained.* Boston: Little, Brown and Co.

———. 1996. *Darwin's dangerous idea: Evolution and the meanings of Life.* New York: Simon and Schuster.

Deputte, B. L. 1982. Duetting in male and female songs of the white-cheeked gibbon. In *Primate communication,* ed. C. T. Snowden, C. H. Brown, and M. R. Petersen, 67–93. New York: Cambridge University Press.

d'Errico, F., C. Henshilwood, M. Vanhaeren, and K. van Niekerk. 2005. *Nassarius kraussianus* shell beads from Blombos Cave: Evidence for symbolic behaviour in the Middle Stone Age. *Journal of Human Evolution* 48(1):3–24.

Dessalles, J.-L. 2008. From metonymy to syntax in the communication of events. *Interaction Studies* 9:51–65.

De Waal, Frans. 1988. The communicative repertoire of captive bonobos compared to that of chimpanzees. *Behavior* 106:183–251.

———. 1992. *The third chimpanzee.* New York: HarperCollins.

———. 1995. Bonobo sex and society. *Scientific American* (March): 82–88.

———. 1997. *Bonobo: The forgotten ape.* Berkeley: University of California Press.

Diamond, Jared M. 1982. Evolution of bowerbirds' bowers: Animal origins of the aesthetic sense. *Nature* 297:99–102.

———. 2005. *Collapse: How Societies Choose to Fail or Succeed.* New York: Viking Books.

Disotell, T. R. 2006. "Chumanzee" evolution: The urge to diverge and merge. *Genome Biology* 7:240.

Dittus, W.P.G. 1984. Toque macaque food calls: semantic communication concerning food distribution in the environment. *Animal Behavior* 32:470–77.

Dobzhansky, T. 1964. Biology, molecular and organismic. *American Zoologist* 4:443–52.

Dogil, G., H. Ackermann, W. Grodd, H. Haider, H. Kamp, J. Mayer, A. Riecker, and D. Wildgruber. 2002. The speaking brain: A tutorial introduction to fMRI experiments in the production of speech, prosody and syntax. *Journal of Neurolinguistics* 15:59–90.

Domínguez-Rodrigo, M., T. R. Pickering, S. Semaw, and M. J. Rogers. 2005. Cutmarked bones from Pliocene archaeological sites at Gona, Afar, Ethiopia: Implications for the function of the world's oldest stone tools. *Journal of Human Evolution* 48(2):109–21.

Dunbar, Robin I. M. 1996. *Grooming, gossip and the evolution of language*. London: Faber and Faber.

Dyer, F. C. and Gould, J. L. 1983. Honey bee navigation. *American Scientist* 71:587–97.

Eldredge, Niles, and Stephen Jay Gould. 1972. Punctuated equilibria: An alternative to phyletic gradualism. In *Models in Paleobiology*, ed. T. M. Schopf, 82–115. San Francisco: Freeman Cooper.

Eliot, T. S. 1998. *The waste land and other poems*. London: Penguin.

Estes, R. D. 1991. *The behavior guide to African mammals*. Berkeley: California University Press.

Evans, P. D., S. L. Gilbert, N. Mekel-Bobrov, E. J. Vallender, J. R. Anderson, L. M. Vaez-Azizi, S. A. Tishkoff, R. R. Hudson, and B. T. Lahn. 2005. Microcephalin, a gene regulating brain size, continues to evolve adaptively in humans. *Science* 309: 1717–20.

Everett, Daniel L. 2005. Cultural constraints on grammar and cognition in Pirahã: Another look at the design features of human language. *Current Anthropology*, August–October.

———. 2007. Cultural constraints on grammar in Pirahã: A reply to Nevins, Pesetsky, and Rodrigues 2007. *LingBuzz*. ling.auf.net/lingbuzz/000427.

Fagan, B. M., and E. L. Van Noten. 1971. *The hunter-gatherers of Gwisho*. Tervuren, Belgium: Musée Royal de l'Afrique Centrale.

Falk, D. 1993. *Braindance*. New York: Henry Holt.

———. 2004. Prelinguistic evolution in early hominins: Whence Motherese? *Behavioral and Brain Sciences* 27:491–503.

Feblot-Augustine, J. 1998. La circulation des matières premières au paléolithique (the movement of raw materials in the Paleolithic). *Journal of Anthropological Research* 54:286–328.

Fitch, W. T., and M. D. Hauser. 1995. Vocal production in non-human primates: Acoustics, physiology, and functional constraints on "honest" advertisement. *American Journal of Primatology* 37:191–219.

Foley, Robert, and Marta M. Lahr. 2005. The origins of modern humans: Insights into speciation. Paper presented at the conference Rethinking the Human Revolution, Cambridge, U.K., September.

Frisch, K. von. 1967. Honeybees: Do they use direction and distance information provided by their dancers? *Science* 158:1072–76.

Frishberg, N. 1987. Home sign. In *The Gallaudet Encyclopedia of Deaf People and Deafness*, vol. 3, 128–31. New York: McGraw-Hill.

Gadagkar, Raghavendra. 1990. Origin and evolution of eusociality: A perspective from studying primitively eusocial wasps. *Journal of Genetics* 69(2):113–25.

Gardner, R. A., and B. T. Gardner. 1969. Teaching sign language to a chimpanzee. *Science* 165:664–72.

———. 1978. Comparative psychology and language acquisition. *Annals of the New York Academy of Sciences* 309:37–76.

Geisemann, T. 2000. Gibbon songs and human music from an evolutionary perspective. In *The origins of music*, ed. N. L. Wallin, B. Merker, and S. Brown, 103–24. Cambridge, Mass.: MIT Press.

Goodall, J., and H. von Lawick. 2000. *In the shadow of man*. New York: Houghton Mifflin.

Goodglass, H., and N. Geschwind. 1976. Language disorders. In *Handbook of Perception: Language and Speech*, vol. 7, ed. E. C. Carterette and M. P. Friedman. New York: Academic Press.

Goodman, C. S. and B. S. Coughlin eds. 2000. The evolution of evo-devo biology. *Proceedings of the National Academy of Sciences* 97:4424–56.

Gould, J. L. 1976. The dance language controversy. *Quarterly Review of Biology* 51(2): 211–44.

Grimshaw, Jane B. 1990. *Argument structure*. Cambridge, Mass.: MIT Press.

Hadingham, Evan. 1980. *Secrets of the ice age*. Northvale, N.J.: Marboro Books.

Hangartner, W. 1969. Trail-laying in the subterranean ant Acanthomyops interjectus. *Journal of Insect Physiology* 15:1–4.

Harnad, S. R., H. D. Steklis, and J. Lancaster, eds. 1976. *Origins and evolution of language and speech*. Annals of the New York Academy of Sciences 280. New York: New York Academy of Sciences.

Hart, Donna, and Robert W. Sussman. 2005. *Man the hunted: Primates, predators and human evolution*. New York: Westview.

Hart, J. Jr., and B. Gordon. 1992. Neural subsystems for object knowledge. *Nature* 359:60–64.

Hauk, O., I. Johnsrude, and F. Pulvermüller. 2004. Somatotopic representation of action words in human motor and premotor cortex. *Neuron* 41(2):301–307.

Hauser, M. D. 1996. *The evolution of communication*. Cambridge, Mass.: MIT Press.

———. 1998. Functional referents and acoustical similarity: Field playback experiments with rhesus monkeys. *Animal Behavior* 55(6):1647–58.

Hauser, M. D., N. Chomsky, and W. T. Fitch. 2002. The language faculty: Who has it, what is it, and how did it evolve? *Science* 298:1569–79.

Hayes, K. J., and C. H. Nissen. 1971. Higher mental functions of a home-raised chimpanzee. In *Behavior of nonhuman primates: Modern research trends*, ed. A. M. Schrier and F. Stollnitz, 59–115. New York: Academic Press.

Hebb, Donald O. 1949. *The organization of behavior*. New York: Wiley.

Heinrich, Bernd. 1991. *Ravens in winter*. London: Vintage.

Herman, L. M. 1986. Cognition and language competencies of bottlenosed dolphins. In *Dolphin cognition and behavior: A comparative approach*, ed. R. J. Schusterman, J. Thomas, and F. G. Wood, 221–51. Hillsdale, N.J.: Lawrence Erlbaum Associates.

Herman, L. M., and P. H. Forestell. 1985. Reporting presence or absence of named objects by a language-trained dolphin. *Neuroscience and Biobehavioral Reviews* 9:667–91.

Herman, L. M., D. G. Richards, and J. P. Wolz. 1984. Comprehension of sentences by bottlenosed dolphins. *Cognition* 16:129–219.

Herrnstein, R. J. 1979. Acquisition, generalization, and discrimination reversal of a natural concept. *Journal of Experimental Psychology: Animal Behavior Processes* 5:116–29.

Hockett, Charles. 1960. The origin of speech. *Scientific American* 203:89–96.

Hockett, Charles, and Robert Ascher. 1964. The human revolution. *Current Anthropology* 5:135–68. Reprinted 1992, *Current Anthropology* 33:7–46.

Holldobler, B. 1971. Recruitment behavior in Camponotus socius. *Zeitschrift fur Vergleichende Physiologie* 75:123–42.

———. 1978. Ethological aspects of chemical communication in ants. *Advances in the Study of Behavior* 8:75–115.

Humphrey, N. K. 1976. The social function of intellect. In *Growing points in ethology*, ed. P.P.G. Bateson and R. A. Hinde, 303–17. Cambridge: Cambridge University Press.

Hunt, Gavin R., and Russell D. Gray. 2003. Diversification and cumulative evolution in New Caledonian crow tool manufacture. *Proceedings of the Royal Society of London B* 270:867–74.

Hurford, James, 2007. *The origins of meaning*. Oxford: Oxford University Press.

Huybregts, Riny. 2006. Development and evolution of human language: An argument from language design. Paper delivered at the Cradle of Language Conference, Stellenbosch, South Africa, November.

Irons, D. B., R. G. Anthony, and J. A. Estes. 1986. Foraging strategies of glaucous-winged gulls in a rocky intertidal community. *Ecology* 67:1460–74.

Isaac, Glyn L. 1977. *Olorgesailie: Archeological studies of a Middle Pleistocene lake basin in Kenya*. Chicago: University of Chicago Press.

Jablonski, Nina G. 2007. A conversation with Nina G. Jablonski. By Claudia Dreifus. *The New York Times*, January 9, 2007.

Jackendoff, Ray. 2002. *Foundations of language: Brain, meaning, grammar, evolution*. Oxford: Oxford University Press.

Jackendoff, Ray, and Steven Pinker. 2005. The nature of the language faculty and its implications for evolution of language (reply to Fitch, Hauser, and Chomsky). *Cognition* 97:211–25.

Jarvis, E. D., and C. V. Mello. 2000. Molecular mapping of brain areas involved in parrot communication, *Journal of Comparative Neurology* 419(1):1–31.

Jellinek, A. 1977. The lower Paleolithic. *Annual Review of Anthropology* 6:1–33.

Jenkins, Lyle. 2000. *Biolinguistics: exploring the biology of language*. Cambridge: Cambridge University Press.

Jesperson, Otto. 1922. *Language: Its nature, development, and origin*. London: Allen and Unwin.

Johansson, Sverker. 2005. *Origins of language: Constrains on hypotheses*. Amsterdam: Benjamins.

Kano, Takayoshi. 1992. *The last ape: Pygmy chimpanzee behavior and ecology*. Stanford, Calif.: Stanford University Press.

Kawai, M. 1965. Newly-acquired pre-cultural behavior of the natural troop of Japanese monkeys on Koshima islet. *Biomedical Life Sciences* 6:1–30.

Kawamura, S. 1959. The process of sub-culture propagation among Japanese macaques. *Primates* 2:43–54.

Klein, Richard. 2002. *The dawn of human culture*. New York: Wiley.

Kohn, M., and S. Mithen. 1999. Handaxes: Products of sexual selection? *Antiquity* 73:518–26.

Kojima, S. 1990. Comparison of auditory functions in the chimpanzee and human. *Folia Primatologia (Basel)* 55(2):62–72.

Krebs, J. R. 1991. Animal communication: Ideas derived from Tinbergen's activities. In *The Tinbergen legacy*, ed. M. S. Dawkins, T. R. Halliday, and R. Dawkins, 60–74. London: Chapman and Hall.

Krunkelsven, E., J. Dupain, L. Van Elsacker, and R. F. Verheyen. 1996. Food calling by captive bonobos (*Pan paniscus*): An experiment. *International Journal of Primatology* 17:207–17.

Laland, Kevin N., and Kim Sterelny. 2006. Seven reasons (not) to neglect niche construction. *Biological Theory* 1 (1):41–43.

Lamberts, Koen, and David Shanks. 1997. *Knowledge, concepts and categories*. Cambridge, Mass.: MIT Press.

Langer, Jonas. 2006. The heterochronic evolution of primate cognitive development. *Evolution* 60:1751–62.

Larick, R. and R. Ciochon. 1996. The African emergence and early Asian dispersals of the genus *Homo. American Scientist*, November–December.

Lee, K. E. 1985. *Earthworms: Their ecology and relation with soil and land use*. London: Academic Press.

Lee, Richard B., and Irven DeVore, eds. 1968. *Man the hunter*. London: Aldine.

Lee, Sang-Hee, and Milford H. Wolpoff. 2003. The pattern of evolution in Pleistocene human brain size. *Paleobiology*, Spring.

Lenneberg, Eric. 1967. *Biological foundations of language*. New York: Wiley.

Lewontin, R. 1983. The organism as subject and object of evolution. *Scientia* 188:65–82.

Lorenz, Konrad. 1937. Biologische Fragestellungen in der Tierpsychologie (Biological questions in animal psychology). *Zeitschrift für Tierpsychologie* 1:24–32.

Lüscher, M. 1961. Air-conditioned termite nests. *Scientific American* 205:138–45.

Macphail, Euan. 1987. The comparative psychology of intelligence. *Behavioral and Brain Sciences* 10:645–95.

Marcus, G. F., and S. E. Fisher. 2003. *FOXP2* in focus: What can genes tell us about speech and language? *Trends in Cognitive Sciences* 7(6):257–62.

Marcus, Gary. 2004. *The Birth of the Mind*. New York: Basic Books.

Marshack, Alexander. 1981. On Paleolithic ochre and the early uses of color and symbol. *Current Anthropology*, 22:188–91.

Martin, A., and Chao, L. L. 2001. Semantic memory and the brain: Structure and processes. *Current Opinion in Neurobiology* 11:194–201.

Masayuki N., E. Kato, Y. Kojima, and N. Itoigawa. 1999. Carrying and washing of grass roots by free-ranging Japanese macaques at Katsuyama. *Folia Primatologica* 69: 35–40.

Maynard Smith, J. 1976. Sexual selection and the handicap principle. *Journal of Theoretical Biology* 57:239–42.

Mayr, Ernst. 1963. *Animal species and evolution*. Cambridge, Mass.: Harvard University Press.

———. 1982. *The growth of biological thought*. Cambridge, Mass.: Belknap Press.

McBrearty, Sally, and Alison F. Brooks. 2000. The revolution that wasn't: A new interpretation of the origin of modern human behavior. *Journal of Human Evolution* 39(5): 453–563.

McCrone, John. 1992. *The ape that spoke*. New York: Avon Books.

McHenry, H. M. 1994. Tempo and mode in human evolution. *Proceedings of the National Academy of Sciences* 91(15):6780–86.

Mellars, P., K. Boyle, O. Bar-Yosef, and C. Stringer, eds. 2007. *Rethinking the human revolution*. McDonald Institute Monographs. Cambridge, U.K.: McDonald Institute for Archaeological Research.

Miles, H. L. 1978. Language acquisition in apes and children. In *Sign language and language acquisition in man and ape*, ed. F.C.C. Peng, 103–20. Boulder, Colo.: Westview Press.

————. 1990. The cognitive foundations for reference in a signing orangutan. In *"Language" and intelligence in monkeys and apes*, ed. S. T. Parker and K. R. Gibson, 511–39. Cambridge: Cambridge University Press.

Miller, Geoffrey F. 1997. Mate choice: From sexual cues to cognitive adaptations. In *Characterizing human psychological adaptations*, Ciba Foundation Symposium 208, ed. G. Cardew, 71–87. New York: Wiley.

————. 2000. *The mating mind: How sexual choice shaped the evolution of human nature*. London: Heinemann.

Mithen, Steven. 2005. *The Singing Neanderthals: The Origin of Language, Music, Mind and Body*. London: Weidenfeld and Nicolson.

Moglich, M., and Holldobler, B. 1975. Communication and orientation during foraging and emigration in the ant *Formica fusca, Journal of Comparative Physiology* 101:275–88.

Morgan, Elaine. 1982. *The aquatic ape*. London: Stein and Day.

Morris, Desmond. 1999. *The naked ape*. New York: Delta.

Monahan, C. M. 1996. New zooarchaeological data from Bed II, Olduvai Gorge, Tanzania: Implications for hominid behavior in the early Pleistocene. *Journal of Human Evolution* 31:93–128.

Müller, F. M., 1970. The science of language. *Nature* 1:256–59.

Müller, J., S. Gaunt, and P. A. Lawrence. 1995. Function of the Polycomb protein is conserved in mice and flies. *Development* 121:2847–52.

Muller-Schwarze, Dietland, and Lixing Sun. 2003. *The beaver: Natural history of a wetland engineer*. Ithaca, N.Y.: Cornell University Press.

Naiman, R. J., C. A. Johnston, and J. C. Kelley. 1988. Alterations of North American streams by beaver. *BioScience* 38:753–62.

Nevins, Andrew Ira, David Pesetsky, and Cilene Rodrigues. 2007. Pirahã exceptionality: A reassessment. LingBuzz. ling.auf.net/lingBuzz.

O'Brien, E. 1981. The projectile capabilities of an Acheulian handaxe from Olorgesailie. *Current Anthropology* 22:76–79.

O'Connell, J. F., K. Hawkes, and N. G. Blurton-Jones. 1988. Hadza scavenging: Implications for Plio-Pleistocene hominid subsistence. *Current Anthropology* 29:356–63.

————. 1999. Grandmothering and the evolution of *Homo erectus*. *Journal of Human Evolution* 36:461–85.

Odling-Smee, F. J., K. N. Laland, and M. W. Feldman. 1996. Niche construction. *American Naturalist* 147:641–48.

————. 2003. *Niche construction: The neglected process in evolution*. Monographs in Population Biology 37. Princeton, N.J.: Princeton University Press.

Odum, Eugene P. 1959. *Fundamentals of ecology*. Los Angeles: Saunders.

Olaf, H., I. Johnsrude, and F. Pulvermuller. 2004. Somatotopic representation of action words in human motor and premotor cortex. *Neuron* 41:301–7.

Packard, Alpheus Spring. 1901. *Lamarck, the founder of evolution: His life and work with translations of his writings on organic evolution*. New York: Longmans, Green and Co.

Panter-Brick, C., R. H. Layton, and P. Rowley-Conwy, eds. 2001. *Hunter-gatherers: An interdisciplinary perspective*. Cambridge: Cambridge University Press.

Patterson, F. G. 1985. Direct-mail funding-solicitation letter. Washington: The Gorilla Foundation.

Patterson, F. G., and E. Linden. 1981. *The education of Koko*. New York: Holt, Rinehart and Winston.

Pearce, John M. 1997. *Cognition in animals*. New York: Psychology Press.

Peirce, Charles S. 1978. *Collected papers, Vol. II: Elements of logic*. Cambridge, Mass.: Belknap Press.

Penn, D., K. J. Holyoak, and D.J. Povinelli. 2008. Darwin's mistake: Explaining the discontinuity between human and nonhuman minds. *Behavioral and Brain Sciences* 31:109–30.

Pepperberg, Irene M. 2000. *The Alex studies: Cognitive and communicative abilities of Grey parrots*. Cambridge, Mass.: Harvard University Press.

———. 2005. An avian perspective on language evolution. In *Language origins: Perspectives on evolution*, ed. M. Tallerman, 239–62. Oxford: Oxford University Press.

Pepys, Samuel. 2000. *The diary of Samuel Pepys: A new and complete transcription*. Berkeley: University of California Press.

Perry, M., R. B. Church, and S. Goldin-Meadow. 1988. Transitional knowledge in the acquistion of concepts. *Cognitive Development* 3:359–400.

Pfungst, Oskar. 1911. *Clever Hans (the horse of Mr. von Osten): A contribution to experimental animal and human psychology*. Trans. C. L. Rahn. New York: Henry Holt.

Pinker, Steven. 1989. *Learnability and cognition: The acquisition of argument structure*. Cambridge, Mass.: MIT Press.

———. 1994. *The language instinct*. New York: Morrow & Co.

Pinker, Steven, and P. Bloom. 1990. Natural language and natural selection. *Behavioral and Brain Sciences* 13:707–84.

Pinker, Steven, and R. Jackendoff. 2005. The faculty of language: What's special about it? *Cognition* 95(2):201–36.

Pitt, W. C., and P. A. Jordan. 1996. Influence of campsites on black bear habitat use and potential impact on caribou restoration. *Restoration Ecology* 4:423–26.

Pollick, A. S., and F.B.M. de Waal. 2007. Ape gestures and language evolution. *Proceedings of the National Academy of Sciences*, 104(19):8184–89.

Povinelli, D. 1996. Chimpanzee theory of mind? The long road to strong inference. In *Theories of theories of mind*, ed. P. Carruthers and P. K. Smith, 293–329. Cambridge: Cambridge University Press.

Premack, David. 1983. The codes of men and beasts. *Behavioral and Brain Sciences* 6:125–67.

———. 2004. Is language the key to human intelligence? *Science* 303:318–20.

Price, Richard. 1976. *The Guiana maroons*. Baltimore: Johns Hopkins University Press.

Pringle, J.W.S. 1975. Insect flight. *Oxford Biology Readers*, vol. 52. London: Oxford University Press.

Pulvermuller, Friedemann. 2002. *The neuroscience of language: On brain circuits of words and serial order*. Cambridge: Cambridge University Press.

Ramus, F., M. D. Hauser, C. Miller, D. Morris, and J. Mehler. 2000. Language discrimination by human newborns and by cotton-top tamarin monkeys. *Science* 288: 349–51.

Reed, Kaye E. 1997. Early hominid evolution and ecological change through the African Plio-Pleistocene. *Journal of Human Evolution* 32:289–322.

Richerson, P. J., and R. Boyd. 2004. *Not by Genes Alone*. Chicago: University of Chicago Press.

Rizzi, Luigi. 2009. Some elements of syntactic computation. In *Biological Foundations and Origins of Syntax*, eds. Derek Bickerton and Eors Szathmary, Strüngmann Forum Reports, vol. 3. Cambridge, Mass.: MIT Press.

Robinson, Stuart. 2006. The phoneme inventory of the Aita dialect of Rotokas. *Oceanic Linguistics* 45:206–209.

Roche, H. et al. 2002. Les sites archaéologiques pio-pléistocènes de la formation de Nachukui, Ouest-Turkana, Kenya: Bilan synthétique 1997–2001 (Plio-Pleistocene archaeological sites in the Nachukui formation, West Turkana, Kenya: Summary of results 1997–2001). *Comptes Rendus Palevol* 2:663–73.

Rumbaugh, Duane M. 1977. *Language Learning by a Chimpanzee: The Lana Project*. New York: Academic Press.

Sainsbury, J. 2006. *John Wilkes: The Lives of a Libertine*. Aldershot, U.K.: Ashgate.

Sane, S. P. 2003. The aerodynamics of insect flight. *Journal of Experimental Biology* 206:4191–4208.

Satchell, J. E. 1983. *Earthworm ecology: From Darwin to vermiculture*. London: Chapman and Hall.

Saussure, Ferdinand de. 2006. *Writings in general linguistics*. Oxford: Oxford University Press.

Savage-Rumbaugh, Susan, and Roger Lewin. 1994. *Kanzi: The ape at the brink of the human mind*. Toronto: Wiley.

Savage-Rumbaugh, Susan, K. McDonald, R. A. Sevcik, W. D. Hopkins, and E. Rupert. 1986. Spontaneous symbol acquisition and communicative use by pygmy chimpanzees (*Pan paniscus*). *Journal of Experimental Psychology: General* 115:211–35.

Savage-Rumbaugh, Susan, J. Murphy, R. A. Sevcik, K. E Brakke, S. L. Williams, and D. M. Rumbaugh. 1993. *Language comprehension in ape and child*. Monographs of the Society for Research in Child Development 233. Chicago: University of Chicago Press.

Schick, K. D. 2001. An examination of Kalambo Falls Acheulean Site B5 from a geoarchaeological perspective. In *Kalambo Falls prehistoric site*, vol. 3, *The earlier cultures: Middle and earlier Stone Age*, ed. J. Desmond Clark, 463–80. Cambridge: Cambridge University Press.

Schick, K. D., and N. Toth. 1993. *Making Silent Stones Speak: Human Evolution and the Dawn of Technology*. New York: Simon and Schuster.

Schmitz, O. J. 1992. Optimal diet selection by white-tailed deer: Balancing reproduction with starvation risk. *Evolutionary Ecology* 6:125–41.

Schusterman, R. J. and K. Krieger. 1984. California sea lions are capable of semantic comprehension. *Psychological Record* 34:3–24.

Semaw, S., M. J. Rogers, J. Quade, P. R. Renne, and R. F. Butler. 2003. 2.6-million-year-old stone tools and associated bones from OGS-6 and OGS-7, Gona, Afar, Ethiopia. *Journal of Human Evolution* 45:165–77.

Shea, J. J. 2006. The origins of lithic projectile point technology: Evidence from Africa, the Levant, and Europe. *Journal of Archaeological Science* 33(6):823–46.

Simoon, F. J. 1969. Primary adult lactose intolerance and the milking habit: A problem in biological and cultural interrelations. *Digestive Diseases and Sciences* 14:819–36.

Slobin, D. L. 2001. Language evolution, acquisition and diachrony: Probing the parallels. In *The evolution of language out of pre-language*, ed. T. Givon and B. F. Malle, 375–92. Amsterdam: Benjamins.

Spoor, F., M. G. Leakey, P. N. Gathego, F. H. Brown, S.C. Antón, I. McDougall, C. Kiarie, F. K. Manthi, and L. N. Leakey. 2007. Implications of new early *Homo* fossils from Ileret, east of Lake Turkana, Kenya. *Nature* 448:688–91.

Stanford, Craig B. 1999. *The hunting ape: Meat-eating and the origins of human behavior.* Princeton, N.J.: Princeton University Press.

Stephens, D. W., and J. R. Krebs. 1986. *Foraging theory.* Princeton, N.J.: Princeton University Press.

Strieder, G. 1994. The vocal control pathways in budgerigars differ from those in songbirds. *Journal of Comparative Neurology* 343(1):35–56.

Stringer, Chris, and Robin McKie. 1996. *African Exodus: The Origins of Modern Humanity.* London: Jonathan Cape.

Sudd, J. H., and N. R. Franks. 1987. *The Behavioral Ecology of Ants.* New York: Chapman & Hall.

Suddendorf, T., and J. Busby. 2003. Mental time travel in animals? *Trends in Cognitive Sciences* 7:391–96.

Számadó, S., and E. Szathmáry. 2006. Selective scenarios for the emergence of natural language. *Trends in Ecology and Evolution* 21(10):555–61.

Tallerman, Maggie. 2007. Did our ancestors speak a holistic protolanguage? *Lingua* 117 (3): 579–604.

———. 2008. Holophrastic protolanguage: Planning, processing, storage, and retrieval. *Interaction Studies* 9(1):84–99.

Tamminga, Carol A. 2000. Procedural memory. *American Journal of Psychiatry* 157:162.

Terrace, H. S. 1979. *Nim: A chimpanzee who learned sign language.* New York: Knopf.

Thor, S. 1995. The genetics of brain development: Conserved programs in flies and mice. *Neuron* 15(5):975–77.

Thorne, A. G., and M. H. Wolpoff. 1992. The multiregional evolution of humans. *Scientific American* 266:28–33.

Tinbergen, Niko. 1963. On aims and methods. *Zeitschrift für Tierpsychologie* 20:410–33.

Tobias, Philip. 1971. *The brain in hominid evolution.* New York: Columbia University Press.

Tomasello, M. 2007. For human eyes only. *The New York Times*, January 13, 2007.

Torigoe, Takashi, and Wataru Takei. 2002. A descriptive analysis of pointing and oral movements in a home sign system. *Sign Language Studies* 2(3):281–95.

Traill, Anthony. 1981. *Phonetic and phonological studies of !Xóõ Bushman.* Ph.D. dissertation, University of the Witwatersrand, Johannesburg.

Tulving, E. 2002. Episodic memory: From mind to brain. *Annual Review of Psychology* 53:1–25.

Turner, Allen. 1997. *Prehistoric mammals.* National Geographic Society.

Ulijaszek, Stanley J. 2002. Human eating behaviour in an evolutionary ecological context. *Proceedings of the Nutrition Society* 61:517–26.

Velasco, J., and Millan, V. H. 1998. Feeding habits of two large insects from a desert stream: *Abedus herberti (Hemiptera: Belostomatidae)* and *Thermonectus marmoratus (Coleoptera: Dytiscidae). Aquatic Insects* 20:85–96.

Voight, B., S. Kudaravalli, X. Wen, and J. K. Pritchard. 2006. A map of recent positive selection in the human genome. *PLoS Biology* 4(3):e72.

Waddington, C. H. 1969. *Towards a theoretical biology.* Vol. 2. Edinburgh, U.K.: Edinburgh University Press.

Walker, A. 1981. Dietary hypotheses and human evolution. *Philosophical Transactions of the Royal Society of London.* B 292:57–64.

Wegener, Alfred. 1924. *The origin of continents and oceans.* London: Methuen.

Weizenbaum, J. 1966. ELIZA, a computer program for the study of natural language communication between man and machine. *Communications of the ACM* 9:25–36.

Wescott, R. W. 1976. Protolinguistics: The study of protolanguages as an aid to glossogonic research. *Annals of the New York Academy of Sciences* 280(1):104–16.

Williams, G. C. 1957. Pleiotropy, natural selection, and the evolution of senescence. *Evolution* 11:398–411.

———. 1992. Gaia, nature worship, and biocentric fallacies. *Quarterly Review of Biology* 67:479–86.

Wilson, E. O. 1962. Chemical communication in the fire ant Solenopsis Saevissima. *Animal Behavior.* 10:134–64.

———. The insect societies. Cambridge, Mass.: Belknap Press of Harvard University Press.

———. 1972. Animal communication. In *The emergence of language: Development and evolution,* ed. W.S.-Y. Wang, 3–15. New York: Freeman and Co.

———. 1980. Caste and division of labor in leaf-cutter ants (Hymenoptera: Formicidae: Atta) *Behavioral Ecology and Sociobiology* 7:157–65.

Wolpoff, Milford H. 1973. Posterior tooth size, body size, and diet in South African gracile australopithecines. *American Journal of Physical Anthropology* 39:376–93.

Worden, R. 1998. The evolution of language from social intelligence. In *Approaches to the evolution of language,* ed. J. R. Hurford, M. Studdert-Kennedy, and C. Knight, 148–68. Cambridge: Cambridge University Press.

Wrangham, Richard, and Dale Peterson. 1998. *Demonic males: Apes and the origin of human violence.* Boston: Houghton Mifflin.

Wray, A. 1998. Protolanguage as a holistic system for social interaction. *Language and Communication* 18:47–67.

———. 2000. Holistic utterances in protolanguage: The link from primates to humans. In *The evolutionary emergence of language: Social function and the origins of linguistic form,* ed. C. Knight, M. Studdert-Kennedy, and J. Hurford, 285–302. Cambridge: Cambridge University Press.

Wynne, Clive D. L. 2004. *Do animals think?* Princeton, N.J.: Princeton University Press.

Zahavi, A. 1975. Mate selection—a selection for a handicap. *Journal of Theoretical Biology* 53:205–14.

Zahavi, A. 1977. The cost of honesty (further remarks on the handicap principle). *Journal of Theoretical Biology* 67:603–605.

Zuberbühler, K. 2002. A syntactic rule in forest monkey communication. *Animal Behavior* 63:293–99.

———. 2005. Linguistic prerequisites in the primate lineage. In *Language origins,* ed. M. Tallerman, 262–82. Oxford: Oxford University Press.

Zuberbühler, K., D. L. Cheney, and R. Seyfarth. 1999. Conceptual semantics in a nonhuman primate. *Journal of Comparative Psychology* 113:33–42.

ACKNOWLEDGMENTS

This book has benefited from discussion and correspondence with a number of scholars, including Michael Arbib, Jill Bowie, Robbins Burling, William Calvin, Noam Chomsky, Tim Crow, Terrence Deacon, Daniel Dennett, Robin Dunbar, Tecumseh Fitch, Tom Givon, Myrna Gopnik, Marc Hauser, James Hurford, Ray Jackendoff, Sverker Johansson, Chris Knight, Steven Mithen, Frederick Newmeyer, Csaba Pléh, Eörs Szathmáry, Maggie Tallerman, and Alison Wray.

I am most grateful to Eörs Szathmáry, Csaba Pléh, and the Collegium Budapest for the four months I spent as a visiting scholar at that institution in 2002, and the interactions with both resident and visiting scholars that took place there.

I thank the then mayor of Barcelona, Joan Clos, for inviting me to take part in Barcelona's Forum 2004, where I first made the acquaintance of niche construction theory, and John Odling-Smee for supplying me with materials relating to the theory and critically reading those sections of the present volume that deal with it.

I am also indebted to Sue Savage-Rumbaugh for an invitation to visit the Great Ape Foundation in Des Moines, Iowa, and for discussion with her and her colleagues there.

Any imperfections that remain are my own responsibility entirely.

INDEX